What Is IQ?

ABOUT THE BOOK

Everyone worries about his IQ. Here is a handbook on intelligence testing to set your mind at rest. Is intelligence inherited or learned? In *What Is IQ?* an authority on intelligence tests exposes the intelligence racket.

There is no evidence that performance on intelligence tests is due to the structure of the brain. There is unequivocal evidence that intelligence tests derive from a tradition of class-consciousness and exaggeration of racial and social differences.

What do intelligence tests measure?
They measure the ability to take intelligence tests.
Who invented intelligence tests?
Dalton in the 19th century. He thought dogs were more intelligent than inferior human beings.
Can intelligence be improved?
There is evidence that people can be trained to excel in intelligence tests.
What is the social purpose of intelligence tests?
To define a race of supermen.
What is their value to us today?
They measure the ability to follow instructions.

ABOUT THE AUTHOR

Carl Liungman lives in Sweden with his wife and two small boys. He has studied philosophy and psychology, has lived in Africa and India.

Aside from his writings on psychiatry and psychology he also writes novels.

His attempts to write this book within a university setting met with official discouragement and he was forced to withdraw and to conduct his research elsewhere. The results, however, were worth while. *What Is IQ?* has been widely acclaimed and has been translated into seven languages.

Carl G. Liungman

What Is IQ?

Intelligence, Heredity and Environment

Translated by Patricia Crampton

GORDON CREMONESI

Designed by Heather Gordon-Cremonesi
Produced by Chris Pye
Set by Preface Ltd, Salisbury
Printed in Great Britain by The Anchor Press Ltd

ISBN 0-86033-003-6

Gordon Cremonesi Ltd
New River House
34 Seymour Road
London N8 OBE
England

Contents

1

PSYCHOLOGY BOOKS MUST BE READ CRITICALLY

The purpose of this book is to describe what intelligence tests are and what results have been achieved by research on intelligence.

Before we begin to make any claims about "intelligence" it is important to know exactly what we mean by it. When we read other people's claims about intelligence, we must know exactly how *they* define the term. For instance, the most widely used psychology textbook in Swedish colleges and technical schools has this to say: "Intelligence quotient score: a measure of the ability to think, according to which 100 is the average . . . intelligence quotients vary, broadly speaking, between 60 and 140." A few pages further on in the book, the text and an accompanying diagram reveal that, on average, the intelligence of Negroes in America borders on imbecility. The textbook is right: most tests in the USA and Canada have resulted in an average IQ of 85 for Negroes, as against 100 for whites. The danger is that a student who has not yet learned to think independently could easily be misled into believing that Negroes have a considerably lower thinking capacity than whites.

Since there are similar average deviations between people who live in the country and people from the big cities and between those who are "only" children and those from large families and between office-workers and drivers, one might be tempted to suppose that farmers, drivers and people with lots of brothers and sisters are less able to think than other people.

We must therefore be absolutely clear as to what it is that Negroes in America have to a lesser degree than whites, farmers less than city-dwellers, drivers less than office-workers and "only" children have more than children from large families. We shall soon see that it is not the ability to think, if the ability to think means the ability to find, without visible activity, solutions to problems of various kinds.

Obviously any assessment of the thinking capacity of one human being in relation to another human being must be based on the peculiar circumstances of his life. This means that any objective comparison of thinking capacity across cultural barriers demands complete familiarity with the language, values, living standards, climate and biological environment of all the cultures concerned.

What do you *mean by intelligence?*

I suspect that most people, when they talk about intelligence or merely read the word, mean an inherited characteristic which "exists" in the brain, and which different people have inherited in different quantities. This characteristic, they believe, decides a person's ability to grasp complicated ideas, the amount he can learn and the quality or clarity of his thinking. High intelligence is generally regarded as necessary to success in life. Since it is inherited from one's ancestors, most people will be so handicapped from birth that they can never reach higher positions in the community. Their heritage is so poor that they will be useless in an occupation which places any great demands on their mental capacities.

This interpretation of the word "intelligence" is based on a thoroughly pessimistic view of humanity. Luckily it is logically unjustified, by the research results assembled from intelligence tests. The more data are collected, the clearer this will become.

2

WHAT IS INTELLIGENCE?

Many people maintain the illusion that there is only one answer to the question "what is intelligence?" They think that since the word "intelligence" exists, and since it is used by psychologists and other scientists as well as by ordinary people, there must be something in every human being which *is* intelligence.

In fact, the word intelligence resembles the word "aggression" in one important respect. Aggression cannot be seen, nor is it possible to point to any definite or indefinite border-line between aggression and any other human attribute. *Aggression does not exist* in the way that the attributes of corpulence, long-leggedness or blue eyes exist. Aggression is a construct which corresponds to the expectations of other people. The statement "John has the attribute of aggression" or "John is aggressive" means "Charles, Fred and Sandra-Jane believe that when John comes up against certain things he begins to hit out."

The same is true of "intelligence". The statement "Emma is very intelligent" means "Mary, Michael and Louise know that if they give Emma a certain type of problem to solve she will tackle it better than most people." The assumption that Emma possesses the attribute of intelligence is based on the fact that other people have previously seen or heard Emma behave in certain ways.

This means that the question "What is intelligence?" can be translated as "What was it that made Emma behave in that way?" or "What was it that made Trevor behave in that way?" and so on.

One way of getting closer to the answer to the question: "What *is* intelligence?" is to turn it into the question "What type of behaviour do the psychologists describe as intelligent?"

The answer will consist of a number of statements of this type: "Mr. Johnson, the psychologist, says that Peter's behaviour is intelligent when Peter does such-and-such in this or that situation."

The next step in the study of what intelligence actually *is* consists in gathering together all the different psychologists' statements and producing a behaviour group which most psychologists call intelligent. Then one can begin to look into what makes people behave in those ways. Intelligence *is* what makes people behave in those ways.

What makes people behave in ways which psychologists call intelligent?

Having decided what types of behaviour can correctly be called intelligent, one may then have a belief, or an intuitive idea, about what it is that makes many people behave in these ways and not in unintelligent ways.

One can also go about it scientifically. In this case a situation is arranged in which people have the choice of behaving in various ways. Then a number of people are put in this situation and observations are made as to which behave intelligently and which do not. After that the differences between the people who behaved intelligently and those who did not are studied.

By tracing these differences and studying them in more detail, a number of more or less distinct factors will emerge which are connected with the fact that people behave intelligently.

Of course, such people are human beings and move about and communicate with the surrounding world in order to behave intelligently. But this is not all: a number of other factors must also be present.

The question "What *is* intelligence?" can now be answered by adding up all the factors which are sufficient to make a person behave intelligently.

Most people will not accept this as an answer to the question. They want an answer which defines a small portion of the human being, preferably something in the structure of the brain, as "the intelligence." If an answer of this type is to have any meaning, then the portion indicated must be regarded as the most important factor of all those which add up to making a human being behave intelligently.

Does it matter whether intelligence is hereditary?

The most important question facing intelligence research workers is whether variations in intelligence are mainly dependent on heredity or are produced by the environment. But many psychologists go round and round the question, like cats circling hedgehogs. Some try to prove that the question itself is silly, others adopt a standpoint. In this book the question will receive the highest attention and the many available research results which may provide an answer to the question will be reviewed.

The question is also important from the point of view of the national economy. If it could be shown that intelligence is principally dependent on heredity, then the intelligence of a population might be increased only by genetic legislation or biochemical interference with genetic factors which we have not yet mastered. For instance, subsidies might be handed out to increase the birth-rate of families of above-average intelligence and deterrents imposed on reproduction by families of below-average intelligence. This would mean that those who were lucky enough to be born of intelligent parents would be further subsidized by income transferred from the less intelligent tax-payers, which would increase social divisions. If we did not take such measures the average level of intelligence of our population would fall steadily, because on average children from large families have lower intelligence and because social groups with the highest intelligence propagate themselves a little more slowly than other social groups.

If, on the other hand, intelligence is conditional on the environment, so that all adults born with normal nervous systems would have the same level of intelligence if they underwent exactly the same experiences, then the community possesses a huge reserve of talent, of which only a small percentage is currently exploited. In this case, a relatively small allocation of public funds for the education of parents and education in general would produce tremendous dividends. For instance, parents might be rewarded for taking courses on how to behave towards their children so as to encourage intellectual development, and all teachers might be specially trained in the art of awakening and developing intellectual capacities in children with normal nervous systems whose intelligence has been handicapped by their home environments.

The prevalent view among psychologists today is that heredity and environment constantly interact and that the two factors consequently cannot be separated, but that genetic heredity sets an upper limit on individuals' intelligence potentials. Close to this limit, an advantageous intellectual environment becomes less important and ultimately cannot help to increase intelligence. Many modern psychologists therefore claim that the question of heredity and environment is becoming less interesting. But this is not so. If the development of the school system went far enough, the "intelligence ceiling" of the entire population would be reached and then the same problems would arise as if intelligence were entirely dependent on heredity.

Are people with high IQs more valuable to the community?

I deliberately avoided defining "intelligence" in the previous section. I wanted to demonstrate that if we replaced "intelligence" by "the results of intelligence tests" in the text, the question under discussion would lose a great deal of its significance.

Economically speaking it is of little importance how much of the population has an IQ score of over 120 (normal student level). The IQ score is not very illuminating about the individual's usefulness to the community. There are those who claim that it reveals more about his ability to do crossword puzzles. Creativity is more likely the attribute of most importance to the community. "Creativity" means the capacity to create in the most general sense, the ability to arrive at new solutions to new problems or to old, unsolved problems, the ability to make things happen in the way one wants them to.

Nevertheless an IQ score appears to have some economic significance, in that a high IQ still indicates a more than usually harmonious, and often intellectual, home environment and a good school education at an early stage. And these factors in turn often mean a useful citizen who will reach a high social position. The apparent economic significance in high IQ scores has led to the notion that if we could bring the average IQ of the whole population up to 120, the majority of the population would consist of harmonious, cultured people with a high level of education and a high capacity to solve intellectual problems. This notion is mistaken. If we invested the necessary resources, we could actually raise the average IQ of any population from 100 to 120 in a very short time. Intelligence tests, as we shall see, can themselves easily train people to solve the types of problems which occur in them. But it is obvious that this would not lead to a dramatic change in people's real-life capacities.

Do intelligence tests measure real intelligence?

The fact is that intelligence-test results say very little about a human being's capacities in life but they say a great deal more about his parents' vocabulary, ambitions and methods of upbringing. This has led many psychologists and others to make statements along these lines: "The intelligence tests now in use are not particularly good for measuring intelligence. We shall have to find new tests which measure real intelligence."

Statements like this show that the idea of a biologically inherited, qualitatively varied mental capacity is deeply rooted. This idea is to some extent a matter of faith, a doctrine. It is always assumed to be true and can never be uprooted by scientific evidence. Every new test which claimed to measure something which was not genetically based would be accused of not measuring "true" or "real" intelligence. New tests would have to be designed forever until some test could be proved to measure an attribute which was genetically based, or until those people who subscribe to the doctrine agreed to accept a more or less precise definition of "real" intelligence.

So in this book I shall use the word "intelligence" in the sense of "what is measured by intelligence tests." This is a precise definition, because psychologists are agreed as to which tests are to be called intelligence tests and which are not.

3
FRANCIS GALTON AND THE FIRST INTELLIGENCE TEST

The age of intellectual competition

Sir Francis Galton was the youngest of a family of four sisters and three brothers in a well-to-do home. His father had inherited an armaments factory, a profitable banking business and the Quaker faith and he left Francis with quite enough money to be able to do what he liked and, presumably, with the urge to find a natural law which would justify his socially privileged position.

Francis Galton is regarded as a genius. He proved that human finger-prints (they were already known at that time) do not change during his life-time, with the result that police everywhere now use finger-prints as one of the most important ways of proving guilt. He made voyages of discovery to then unknown parts of the world. He introduced the words "anticyclone" and "high pressure" and gave meteorology a hefty push forward. Above all, he introduced methods for the numerical classification of various physiological and mental attributes.

The well-known American professor of psychology, Lewis Terman, has drawn up long lists of the IQ scores of famous men. He evaluated their intelligence from what he could elicit from their biographies. Lists of this kind tend to occur in manuals of psychology of the type used in introductory psychology courses at university. According to Terman, Galton had an IQ of 200, the highest of all those he studied. Galton's famous cousin Charles Darwin had to be content with 135. When we get IQ figures as high as 200, they are of little value as material for comparison with more average individuals. It is better to regard Galton's score of 200 as a kind of definition of "intelligence."

As an example of his genius, here is a letter which Francis Galton wrote on *the day before his fifth birthday* to his sister Adèle, who was something of an invalid and had become the person mainly concerned in encouraging Francis to outshine his contemporaries in intellectual achievements:

> Dear Adèle,
> I am four years old, and I can read any English book. I can say all the Latin substantives, adjectives, and active verbs, besides 52 lines of Latin poetry. I can cast up any sum in addition and multiply by 2, 3, 4, 5, 6, 7, 8, [9], 10, [11]. I can read French a little and I know the clock.
> Francis Galton Feb[r]uary 15, 1827.

Francis probably added the words in brackets later. The alterations were made with a razor-blade and ink. Note that he emphasizes that he is four years old when he can do all these things.

Galton did great work in the service of science. Nevertheless, he was a victim of the weaknesses of his time. He was unconscious of the relativity of value judgments such as "beautiful"; this is shown by the map he made of the geographical distribution of "beauty" in England. He also had strange ideas about the intellectual gulf between human beings and animals. In his famous book "Hereditary Genius" he claims that an intelligent dog is superior to an imbecilic human in mental capacity and memory. He was absolutely convinced of the exclusively hereditary nature of genius and of the great mental superiority of certain races. In the same book he writes: "To conclude, the range of mental power between – I will not say the highest Caucasian and the lowest savage – between the greatest and least English intellects, is enormous."

Francis Galton wanted to be a doctor and studied medicine when he left school. He abandoned his medical studies for various reasons and then travelled in Eastern

Europe and Turkey. On his return he began to study mathematics at Cambridge. Before he could complete the course he had started to take part in the momentous intellectual trial of strength which all mathematics students at Cambridge had to undergo at the end of their studies, he fell sick. Perhaps a psychoanalyst would find interesting implications in the fact that thereafter Galton showed a great propensity for numerical relationships and was always busy working out the heights of mountain peaks and the number of flowers on every bush he passed, instead of allowing himself to be conscious of natural beauty. A Swedish psychologist writes: "He was obsessed with measuring: the pitch of notes, the lengths of arms, the diameters of seeds, the speed of mental connections, the capacity of people to listen to a lecture without falling asleep. There was no limit to the measurable."

The fact is that Galton based his assumptions on the order in which mathematical students at Cambridge were graded: in "Hereditary Genius" he claimed there were enormous biologically inherited differences in the quality of understanding between different individuals. According to the author quoted above: "When Galton looked through these lists [the results of mathematics exams] he was struck by the peculiar fact that there was such a large gap between the highest and the lowest marks." Unfortunately he was never struck by the fact that the wide differences in marks corresponded to a convention among the examiners – the convention as to how many marks should be allotted to a given degree of elegance in a solution or to a number of correct solutions. In other words, he treated a conventionally imposed numerical value as if it were an objective reality which corresponded to an equally great divergence in physiological capacity.

To return to Galton's life, before he set out on these lines his father died in the same year as he ended his mathematical studies and he inherited a fortune which made him independent. He then travelled widely in Egypt and the Sudan. When he came home, according to the "Dictionary of National Biography," he devoted himself mainly to sport for five years. He then personally financed an expedition to South-West Equatorial Africa. After his return from this expedition he applied himself to meteorology and shed light on the operation of high pressure.

In 1859 his cousin Charles Darwin had published the famous work which explained the exclusive importance of biological heredity to the existence of new species. Galton claimed that even before the "Origin of Species" appeared in 1859 he had a number of ideas brewing on heredity. He began work immediately on a study of hereditary genius. Ten years after Darwin's book, Galton published "Hereditary Genius" in which he reported the results of a study covering 977 outstanding men (no women) from a total of 300 families. His aim was to prove that genius was exclusively based on heredity *in the human species*, in the same way as the differences *between species* are the result of heredity.

In his choice of subjects for investigation he proceeded on the assumption that reputation is a measure of natural ability. "By reputation I mean the opinion of contemporaries, revised by posterity – the favourable result of a critical analysis of each man's character, by many biographers . . .

"By natural ability I mean those qualities of intellect and disposition which urge and qualify a man to perform acts that lead to reputation." He reasoned that the fact that "to every 10 illustrious men *who have any eminent relations at all*, we find 3 or 4 eminent fathers, 4 or 5 eminent brothers, and 5 or 6 eminent sons . . . " proved the biological heredity of genius.

Galton's belief in the importance of biological heredity was almost religious in character. ". . . A man of genius . . . will display an insight into new conditions and a power of dealing with them which even his most intimate friends were unprepared to accredit him. Many a presumptuous fool has mistaken indifference and neglect for incapacity; and in trying to throw a man of genius on ground where he was unprepared for attack, has himself received a most severe and unexpected fall. I am sure that no one who has had the privilege of mixing in the society of the abler men of any great capital, or who is acquainted with the biographies of the heroes of history, can doubt the existence of grand human animals, of natures preeminently noble, of individuals born to be kings of men."

Francis Galton was the creator of the attribute called "intelligence", in as much as the factors which make a man eminent had not previously been assembled under one heading. Most of the content of the modern concept of "intelligence" was pioneered by Galton.

In the following quotation he approaches an idea which was much discussed in the 1930s and 1940s, when the psychologists discovered that, on average, country dwellers achieved lower scores in intelligence tests than townsmen. The passage quoted may have given rise to the now generally abandoned theory of migration, according to which the most intelligent of the agricultural population made their way to the towns and the stupidest stayed on the land:

> "The meaning of the word 'mediocrity' admits of little doubt. It defines the standard of intellectual power found in most provincial gatherings, because the attractions of a more stirring life in the metropolis and elsewhere, are apt to draw away the abler classes of men, and the silly and the imbecile do not take a part in the gatherings. Hence, the residuum that forms the bulk of the general society of small provincial places is commonly very pure in its mediocrity."

After showing that the various degrees of mental capacity, from genius through mediocrity to idiocy, do exist, Galton goes on to demonstrate that they are biological entities:

> "This law of deviation from an average . . . if [it] be the case for stature, will also be true as regards every other physical feature – as circumference of head, size of brain, weight of grey matter, number of brain fibres, etc., and thence, by a step on which no physiologist will hesitate, as regards mental capacity."

Galton's conclusion in "Hereditary Genius" is still one of the cornerstones of differential psychology, on which normally distributed intelligence test results and the term

IQ are based. If the stupidest idiot has an IQ score of 0 and the average man has an IQ of 100, then the greatest genius will have an IQ score of 200. Galton in fact assumed that the most gifted individuals were as far above the average as the idiot is below the average. "Hence we arrive at the undeniable, but unexpected conclusion, that eminently gifted men are raised as much above mediocrity as idiots are depressed below it, a fact that is calculated to considerably enlarge our ideas of the enormous differences of intellectual gifts between man and man."

Despite his conclusions, often based on shaky logic, as to the hereditary nature of genius, despite his predilection for magnifying the differences between people's ability, despite his dislike for the idea that environment had any importance in the development of individual capacity, Galton was a clear-sighted thinker in other ways. One of his many activities was to arrange for the distribution of a questionnaire on which people were to record their imaginative powers. The results of this investigation are interesting. The form contained questions on whether and in what way the subject could imagine the day's breakfast, whether his "inner eye" could conjure up colour and contours, distance and detail. To his surprise, Galton found a clear difference between scientists and "ordinary people." In "Inquiries into the Human Faculty and its Development" (1883), he writes: "My own conclusion is that an over-clear perception of sharp mental pictures is antagonistic to the development of habits of highly generalized and abstract thought, especially when the steps in thinking proceed with the help of words as symbols, and that if the capacity to see pictures is possessed by men who think hard, so is the tendency to get lost because it is not made use of."

If it is correct, Galton's conclusion means that if we define intelligence as the ability to think, and if we consider that the ability to think means the power to manipulate mental images of former moments of reality, then average individuals possess it as much as scientists. The difference in intelligence must then lie in the fact that scientists think in conceptual symbols (words, figures, etc.) while average men think in terms of pictures that more closely resemble immediate reality, which consists to only a limited extent of words, letters and figures and other mental creations of a more formally defined nature.

Francis Galton invented the concept of correlation, which is another of the cornerstones of intelligence research and differential psychology in general. In 1888 he presented his findings to the Royal Academy. Two interchangeable organs are said to be co-related when a change in one is generally accompanied by a corresponding change in the other. Since a person with long arms generally has long legs, we say that the length of the human arm is closely correlated with the length of the human leg. As a measure of the degree of correlation, we take the so-called correlation coefficient which is said to be infinitely close to 1 if each individual in a group has exactly the same proportions between two different attributes as every other individual in the group. In a group of 12 ordinary children, in which the first child was 1 year old, the next 2, etc., the correlation between the attributes of height and age would approach 1 and would lie in the vicinity of 0.90 and above, depending on the methods used to calculate the correlation. The correlation coefficient for two attributes would lie in the vicinity of 0 if the two attributes were unconnected. In a group of adults between the ages of 18 and 28, the attributes of height and age would have an approximate correlation of 0.00.

At the time of the International Fair in London in 1884 Galton opened a measuring laboratory, which was then kept open several years afterward. There Galton and his colleagues measured differences in the sensory acuity in various individuals, their ability to hear high notes, to distinguish various shades of colour, and their external physical characteristics. Above all he tested the capacity to distinguish between very similar sensory impressions. Having discovered that mental defectives were less able to distinguish between different sensory impressions than normal individuals, he concluded that men of genius must be more able to distinguish between sensory impressions than the average man – a conclusion which he then thought he had confirmed.

Galton was a versatile man. His interests ranged over a

wide field, from spiritualism to mental disease. In order to gain some insight into the genesis of fantasies in mental illness, he decided on one of his morning walks to regard everything he met as a secret spy. The experiment succeeded beyond all expectations. "By the time I reached Piccadilly," he wrote, "every horse at every cab-rank was watching me, either directly or covertly." This self-induced delusion of persecution lasted for eight or nine hours and could be easily revived, even two or three months later.

Towards the end of his life Galton's interest focussed more sharply on eugenics. The individual variation in mental capacity which he thought he had proved to be genetically caused were transferred to nations and races. He espoused the cause of hereditary and racial hygiene and stressed the importance of applying genetic laws deliberately to control human evolution. In his will he left the whole of his fortune to establish a Chair in Eugenics and a Eugenic Institute at London University. The first holder of the Chair was Karl Pearson, one of his pupils, now best known for the development and refinement of correlation statistics.

Francis Galton married the daughter of a cathedral provost, previously headmaster of Harrow. Sadly enough, they never had any children. Otherwise, they might have provided living proof of Galton's theory of the hereditary nature of genius.

Galton had several famous pupils. One of these was Cyril Burt, the English psychologist, who considered that available research results proved that up to 77% of variations in intelligence were genetically based. He was one of the few who had the courage to give direct numerical expression to the biological inheritance of genius. Another pupil was the American Cattell, who introduced into the USA the psycho-physical methods of measurement he had learned in Galton's laboratory.

The first intelligence test

Although Francis Galton was the real pioneer of the whole science which is now called intelligence research, it was not he who designed the first tests to be called "intelligence tests." Galton died in 1911, but the first intelligence test appeared in 1905. Its originators were two French scientists, Dr. Alfred Binet and Dr. Théodore Simon. Extensions and revisions of the problems they composed are still in use and, as Stanford-Binet-60, are the most highly-regarded intelligence tests known today.

Binet and Simon were commissioned by a committee under the French Ministry of Education to devise a method of sorting out the children whose mental defects would prevent them from making use of normal educational facilities. A special school had recently been set up for such children.

The interesting point here is *why special methods were needed* in order to decide which children were mentally defective. It was, after all, blindingly obvious that teachers would send to the new school such pupils as were unable to keep up with their classes. Why were special methods needed, when the committee had expressly stated that candidates for the new school were those who could not take advantage of normal education?

The real reason for adopting special methods, rather than accepting the teachers' assessment of their pupils, may have been that to go to a special school bore a social stigma and that the teachers did not have sufficient authority for the community to accept their judgment as to who should be exposed to this stigma.

Be that as it may, Binet and Simon drew up a series of tests which were tried out on a small group of children and which covered an age range from 3 to maturity. In their definition of normality they announced that great weight was given to the power of judgment, whereas memory was regarded as less important. Examples of tests for two different ages (from the revised test scale of 1908) are given below.

A 7 year-old was regarded as normally intelligent if he could perform all or all but one of the following types of task:

point out objects taken out of a picture,
give the number of his fingers,

copy a written sentence,
copy a triangle and a rhombus,
repeat in the correct order five figures read out to him,
describe the content of a picture,
count up 13 sous (based on French coinage),
give the names of the four types of coin used in the above problem.

An 11 year-old was regarded as normal if he could perform all or all but one of the following types of task:

criticize absurd sentences,
make up a sentence containing three given words,
count up at least 60 words in three minutes,
define a number of given abstract terms,
make a correct sentence out of muddled words.

Binet and Simon thought that "nearly all the phenomena with which psychology concerns itself are phenomena of intelligence; sensation, perception are intellectual manifestations as much as reasoning." To Binet and Simon the French word "intelligence" meant something other than what we mean by "intelligence" today. But they selected a fundamental part of intelligence and that was the part they were trying to measure with their tests. They wrote, "It seems to us that in intelligence there is a fundamental faculty, the alteration or the lack of which is of the utmost importance for practical life. This faculty is judgment, otherwise called good sense, practical sense, initiative, the faculty of adapting oneself to circumstances. To judge well, to understand well, to reason well, these are the essential activities of intelligence . . . The rest of the intellectual capacities seem of little importance in comparison with judgment."

Intelligence was invented, not discovered

Galton, Binet and Simon may be said to have created intelligence. After all, intelligence is a postulated entity which cannot be directly observed. All reasonable statements about intelligence are based on observations of the behaviour of a number of people in different situations.

In order to mean approximately the same thing when we speak of intelligent behaviour, we must have a common idea of what intelligence is. Binet and Simon created the methods (or type of methods) for determining the sphere of behaviour which was to define the new attribute. It was Galton who connected it with the new view of man initiated by his cousin Charles Darwin's discoveries and gave it its bearing on biological survival.

The Binet-Simon test came to the USA, was translated by psychologists and led to a proliferation of similar tests. One of the first and in many respects the most important of these revisions was the one undertaken by Professor Lewis Terman at Stanford University in 1916. This test was in turn revised by Terman with the help of Professor Maude A. Merrill in 1937. The 1937 Stanford-Binet test was used until the 1960s. Terman and Merrill revised the test in 1960 and this last revision is known as Stanford-Binet-60.

After 1911 the further development of the concept of intelligence lay almost entirely in the hands of American psychologists, and does still today. What psychologists elsewhere mean by intelligence is basically defined from data which scientists in the USA have collected in an enormous number of tests and statistical analyses. Broadly speaking, it is only in the matter of heredity and environment and the results of factor analysis that differences of opinion exist between psychologists in the various advanced countries. British psychologists are more ready to follow Galton and generally take the view that variations in intelligence in different individuals are primarily genetically based, whereas American psychologists have a tendency to attribute at least as much importance to the environment as to heredity.

Two important milestones in the development of the concept of intelligence are the Army Alpha and AGCT. In 1917, during World War I, the first big group-tests were launched: Army Alpha and Army Beta. They were administered to almost two million men. During World

War II a similar test battery, the Army General Classification Test, was administered to almost twelve million soldiers. All these tests were subject to a time limit.

4

CORRELATIONS

[This chapter may look laborious, but in order to understand the significance of the correlation figures, it will have to be read.]

Correlation statistics were invented by Francis Galton and developed by his pupils, primarily by Karl Pearson. Expressing correlation coefficients is probably the test psychologist's most important procedure.

We say that the height of a human being is highly correlated with the length of his index finger, which means that most tall people have index fingers which are longer than other people's and most short people have index fingers which are shorter than other people's. The correlation coefficient is 0.70 for the attributes of height and index-finger length, according to Galton's calculations.

Correlation coefficients vary between 0.00, when there is no definite connection between two attributes, and 0.99 when the two attributes correspond fully. In order to understand what, for instance, a correlation coefficient of 0.92 means, it should be seen on a graph. Figure 1 is a graphic representation of the correlation 0.92. This correlation is one of the highest which occurs in intelligence research. If you give an intelligence test to a group of identical (monozygotic) twins, who have grown up together, then the correlation between the IQ of one twin and his twin brother (or her twin sister) is often 0.92 (a more representative figure in this connection is 0.89).

If you look at the diagram in Figure 1 you will see that the correlation 0.92 can be said to mean that if one identical twin, Peter, has an IQ rating of 105, then his twin brother Paul will undoubtedly have an IQ rating between 90 and 115.

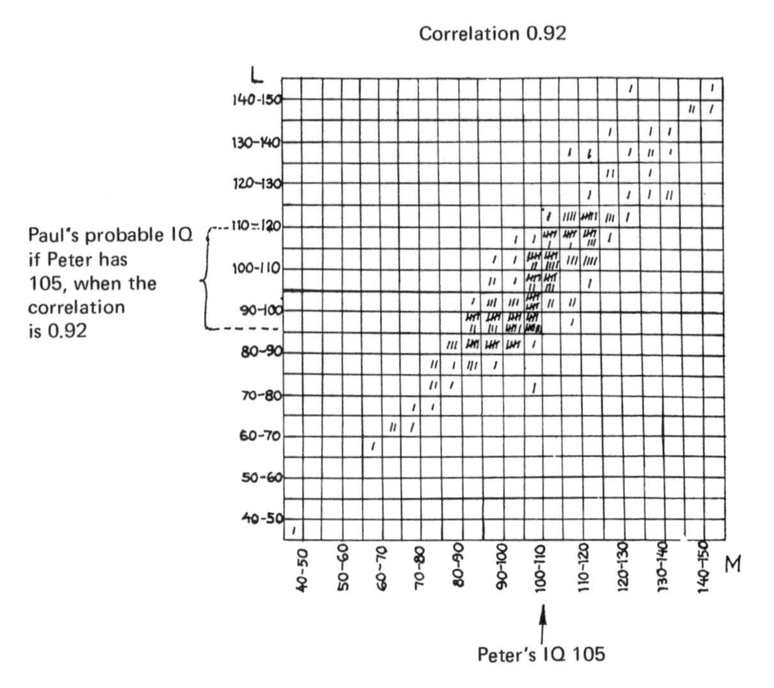

Figure 1. Each of the short strokes in the diagram represents a 7 year-old. The position of the stroke shows how many points the 7 year-old obtained when tested with parallel Form M and how many points he obtained when tested with the other parallel Form L from Stanford-Binet 1937. The coefficient 0.92 shows the reliability of the test for 7 year-olds.

If we know that the correlation between the IQ of identical twins who grow up in the same home is 0.92 and if we know the IQ of one twin, Peter, then we can predict the approximate test results of his twin brother Paul.

If we try to understand what the correlation 0.92 for identical twins means, we have to imagine that each little stroke in the diagram represents a pair of twins. We get the IQ of one twin by following the horizontal line from the little stroke to the left, and reading off the figure at the end. We get his twin brother's (her twin sister's) IQ by following one of the vertical lines down to the scale along the bottom and reading that off.

As an example of a low correlation we will take the IQ

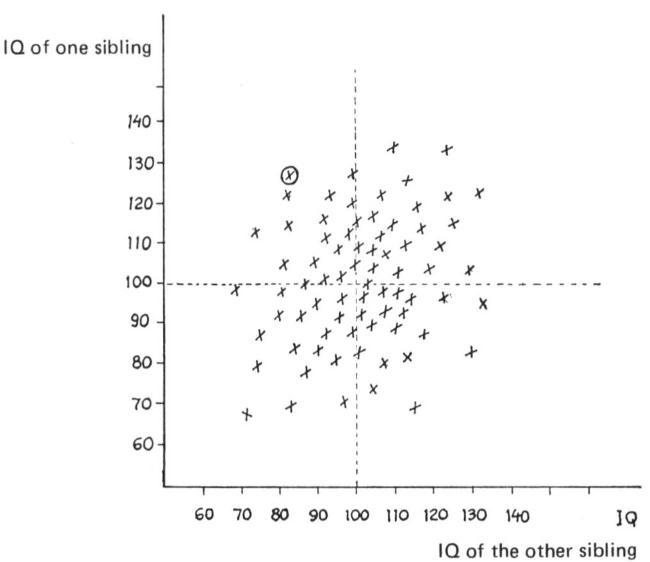

Figure 2. The relationship between the IQ of 76 individuals and 76 of their siblings raised in other families. Each cross represents a pair of siblings.

scores of siblings who have been separated and grown up in different homes. In Figure 2 we can see the results of a study of 76 individuals and 76 of their siblings brought up in different families.

We can see that the intelligence of siblings does not correspond. Each cross on the diagram represents a pair of siblings. It is not at all unusual for one to have a very low IQ and one to have a very high IQ. The cross ringed on the diagram is an example. It indicates a pair of siblings, of whom one had an IQ of only about 80, while the other had a score of nearly 130.

Figure 2 shows that the correlation 0.25 is so low that there is scarcely any observable tendency for the IQs of siblings to be similar. Where correlation coefficients occur in a psychological connection one has to remember that coefficients under about 0.60 indicate scarcely observable tendencies. Individual predictions of the type "John has an

IQ rating of 120. Therefore his sister Cecilia also has 120, since the correlation between the IQ of siblings is 0.50" are quite meaningless. The chances that Cecilia's IQ score is different from random selection of any other person off the street are very low. The possibility that Cecilia's IQ is more like John's IQ than it is like that of the man in the street is approximately fifty-fifty.

The correlation coefficient 0.50 applies to the relationship between the IQ's of husband and wife. It also applies to the IQ scores of siblings who have grown up together and the IQ scores of mothers and their children. Figure 3 shows what a correlation coefficient of 0.50 means to the common variation. We see that there is a tendency for any one individual to have an IQ which is not too unlike any other's.

Figure 3 shows that mothers with an IQ over 100 often have children with an IQ over 100, but in approximately one-third of the cases their children have an IQ of less than 100. This third is marked by a ring round the cross.

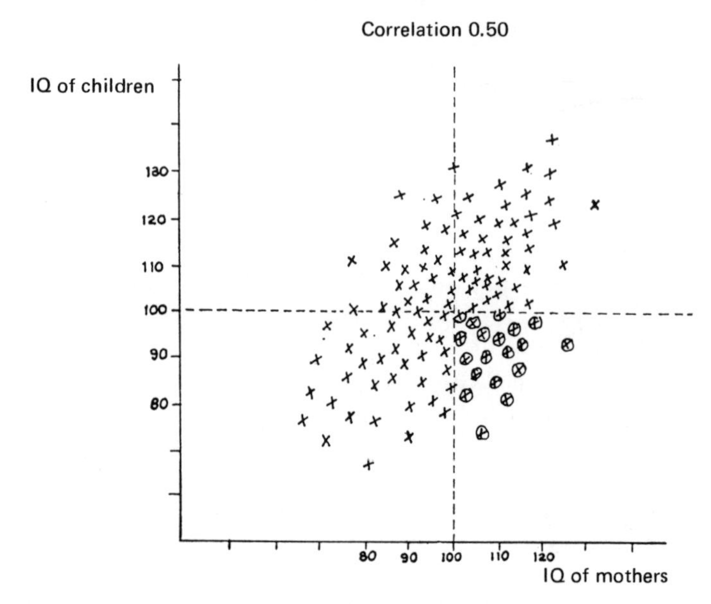

Figure 3. The relationship between the IQ of 118 mothers and 118 of their children. Each cross in the diagram represents a mother-child pair.

Correlation 0.77

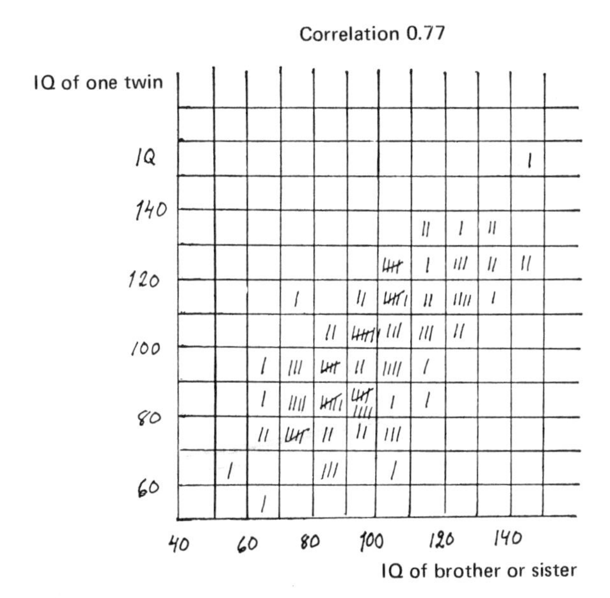

Figure 4. The relationship between the IQ of 99 monozygotic twins separated at approximately one year and raised in different environments. Each stroke in the diagram represents a twin pair.

As a final example we will take the correlation coefficient 0.77. This expresses the relationship between the IQ of identical twins who have been separated, adopted by different families and brought up in different homes. Figure 4 shows that if one twin has an IQ rating of 100 the other twin may have any IQ rating between 70 and 120. But at the same time there is a clear tendency for an above-average intelligent twin to have an above-average intelligent twin brother or sister and in general a twin will be of below-average intelligence if his twin brother or sister is so too.

It is easy to be influenced by correlation figures. The correlation coefficient 0.50, for instance, describes the relationship between the IQ of husband and wife (people most often choose partners not too unlike themselves in social background and general intellectual level). This does not mean that if the man has an IQ rating of 115, in 50% of all cases his wife also has an IQ rating of 115. What it actually means is that his wife has an IQ between 70 and

130 rather more often than would have been expected from a mere random selection.

It is important not to confuse causal relations with correlations. The fact that there is a correlation of 0.35 between the parental tendency to beat children and the criminality of the children does not necessarily mean that beating is the cause of children committing crimes. It may equally well be that the child does a lot of naughty things and that is why his parents beat him. It may also be that parents of criminal children have some attribute which, on the one hand, makes them use force against their own children and, on the other hand, is handed down to the children by heredity and makes him commit crimes. In other words a correlation coefficient in itself says nothing about causal connections.

Here is an example: during the Italian campaign in World War II it was found at Allied Propaganda Headquarters that the correlation between the number of leaflets dropped behind the German lines and the area of territory captured from the Germans was high. Some interpreted this to mean that larger numbers of leaflets should be printed and dropped, in order to win still more territory from the Germans. In fact, larger numbers of leaflets than usual were always dropped behind the German lines just before a major offensive, on the orders of the High Command. The cause of the territorial gains was much more probably the use of manpower and weapons than the use of leaflets. If the Germans had been observant, the frequency of leaflet drops could, and probably to some extent did, help them to work out where and when major assaults were to be mounted. The real connection between dropping leaflets and territorial gains was probably that dropping leaflets caused the gains to be somewhat less than they might otherwise have been.

Where we find negative correlation coefficients, this means that the relationship has turned back to front. The correlation of –0.30 between the degree of strictness in the home and the childrens' IQ means that children from homes where parents are very strict have a tendency to be less intelligent than the children from homes where parents are more liberal in their views and attitudes. Negative correlation coefficients are a function of the way we classify the attributes being correlated. The opposite to strictness is flexibility. One might just as well say: the correlation between the parents' degree of flexibility and their childrens' IQ is positive and is about 0.30.

Correlation coefficients from groups where the individuals differ widely cannot be compared with coefficients from groups where the individuals are more alike.

It is important to recognize that one cannot quite simply compare all the correlation coefficients from studies of two different groups of people. We first have to know how many average individuals there are in each group and how great the difference is between the best and worst individuals in each group. The correlation between schoolchildrens' capacity to count and their IQ is approximately 0.70. But if we take a less heterogeneous group of schoolchildren, e.g. technical school pupils, and work out the correlation between their IQ and their capacity to count, using the same tests as are given to all pupils, then the correlation will be considerably lower, about 0.40. This is because among schoolchildren in general there are some who go to remedial classes and some who are sufficiently able to get into academic classes, in addition to those who are going on to technical school. Students who attend technical school, in other words, are a more homogeneous group than schoolchildren in general as far as IQ and mathematical test results are concerned.

Figure 5 shows the correlation between IQ and mathematical test results for a group of randomly selected schoolchildren. Each cross stands for a pupil who is about to enter an academic course, each little ring means a pupil who is going to technical school and each line means a pupil who is going on to some other type of activity. For the whole group we see a clear tendency for those who have the highest points in the mathematical tests to have the highest IQ score as well.

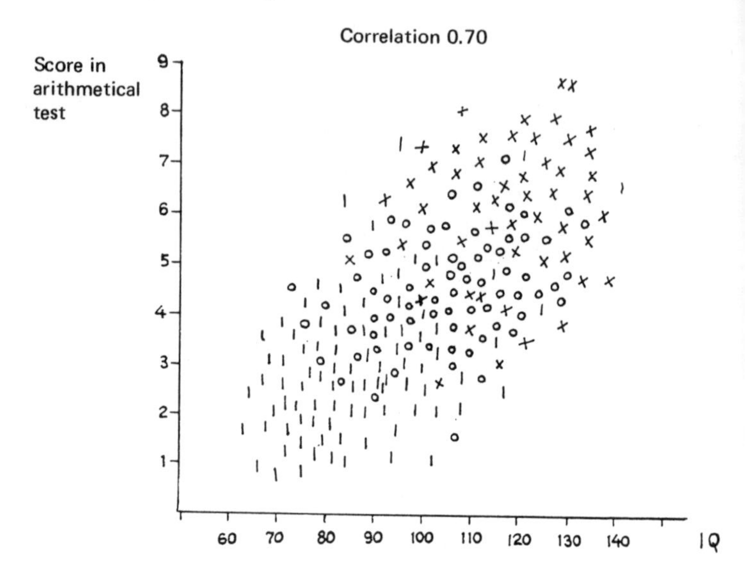

Figure 5. The relationship between IQ and scores gained in an arithmetical test administered to 230 randomly selected schoolchildren. Each cross represents a pupil about to begin on academic studies, each ring represents a pupil about to begin at technical school and each stroke represents a pupil about to become an industrial worker, etc.

Figure 6 is the same as Figure 5, with the difference that the little crosses and lines representing children who were attending academic courses and other training schools have been removed. Only the little rings representing pupils who were going to technical school are left.

The remaining group of 80 technical school pupils is a much more homogeneous group than the group of 230 schoolchildren of all types. In other words, the differences in IQ and mathematical test results between the best and the worst are smaller among the technical school candidates than among the 230 schoolchildren as a whole.

In Figure 5 we can see that all the signs representing schoolchildren of different types lie in a length-wise pattern which rises diagonally to the right of the diagram. The fact that this length-wise pattern lies diagonally is a sign that the correlation is high.

When we take away all schoolchildren except the

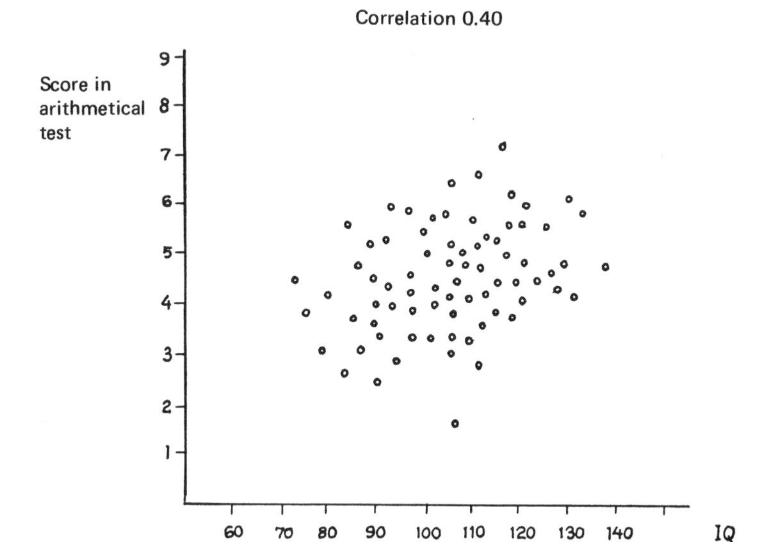

Figure 6. The relationship between scores in the arithmetical test and the IQ of 80 of the 230 schoolchildren in Figure 5. Each ring represents a pupil about to begin at technical school.

technical school candidates we get Figure 6. In this Figure we cannot see any length-wise pattern to speak of. In fact, the little rings lie in an amorphous mass. This is a sign of a low correlation.

Generally speaking, the more homogeneous the group, the lower the correlation. If we take a group of students and give them two different intelligence tests there will be a low correlation between the results. An individual who is above average in one test may often be below average in the other. But if we take a heterogeneous group, for instance a group of backward children, bus-drivers, office workers and university lecturers and give this group two intelligence tests, there will be a high correlation between the results. All the backward subjects will have below-average results in both tests and the University lecturers will have above-average results in both.

Owing to the factors discussed above, it cannot simply be assumed that, for instance, a correlation coefficient of 0.70 from *one* study of twins means the same degree of variation as a correlation of 0.70 from *another* study. The groups may be of different homogeneity.

Dr. Alice Leahy investigated the IQs of 194 children and their adoptive mothers and found a correlation of 0.24. The correlation between the IQs of the adoptive mothers and 194 of their own children was about 0.51. It is easy to conclude that the relationship between the IQs of adopted children and adoptive mothers is only half as great as that between natural mothers and their children. Apart from the fact that such expressions as "twice as great" and "half as great" are meaningless applied to correlations, the figures are in any case misleading. Proportionally speaking, there were considerably more children with very high and very low IQs among the mothers' own children than among their adopted children. The adopted children formed a more homogeneous group, as regards their IQs.

This was the result of several factors. On the one hand, the most backward children at the children's home were seldom adopted, because prospective parents tend to select children who are bright and forthcoming and appear "good". On the other hand, life in a children's home for the first 10–20 months of life affects children in such a way that on average they do not reach the same IQ level as other children later in life (the environment is generally sterile; the care is relatively impersonal and unemotional). The position of an adopted child in a family home also implies a particular psychological climate which is quite similar in different adoptive homes.

If the IQ scores of adopted children were worked out so that the group of adopted children was as heterogeneous as that of "natural" children, the correlation between the adopted children's IQs and the mothers' IQs would be higher.

One measure of the homogeneity of a group is the so-called standard deviation, abbreviated to SD. In Dr. Levy's study the SD was 12.5 for the adopted child group, as against 15.4 for the group of natural children. Figure 7 shows the IQ distribution in a homogeneous group. Compare this with the IQ distribution in a heterogeneous group in Figure 8.

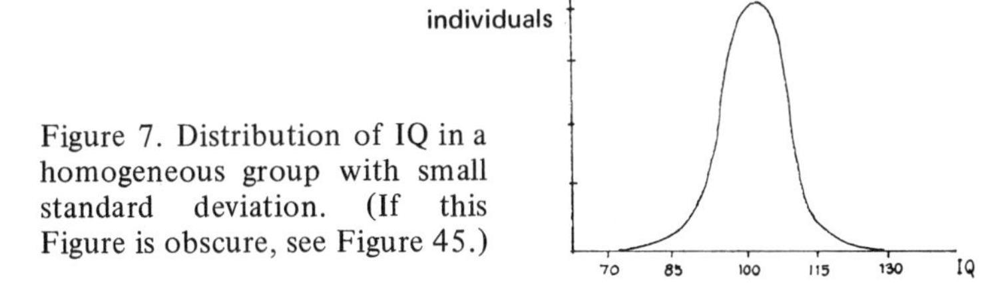

Figure 7. Distribution of IQ in a homogeneous group with small standard deviation. (If this Figure is obscure, see Figure 45.)

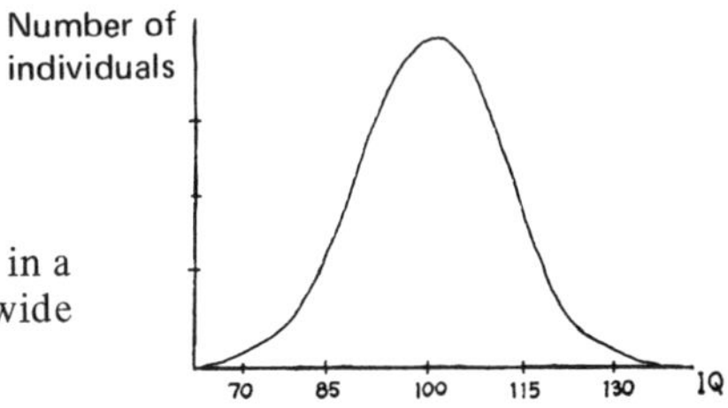

Figure 8. Distribution of IQ in a heterogeneous group with wide standard deviation.

5

TYPES OF INTELLIGENCE TEST

Intelligence tests are usually divided into group tests and individual tests. The group test consists of a form giving a large number of different problems. The problems almost always have to be solved within a specific length of time. Group tests are generally used in the Forces and in schools. The military group test which Swedish recruits have to take when they sign on takes only about half an hour. Other group intelligence tests take longer. A Swedish test developed in Uppsala in the 1940s consisted of two forms which together took about three hours to complete. Here are some examples of problems from the Uppsala test. The problems are quite representative of problems in group tests and in general.

Figures 9 and 10 are two examples of non-verbal problems.

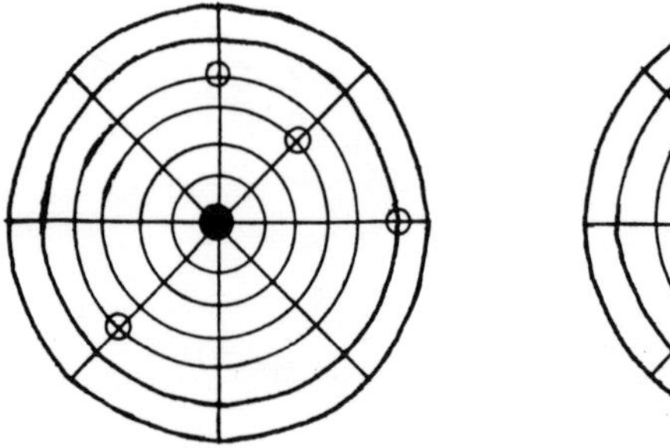

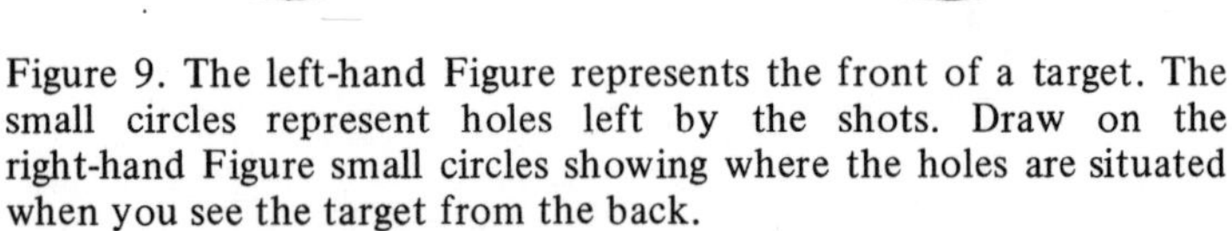

Figure 9. The left-hand Figure represents the front of a target. The small circles represent holes left by the shots. Draw on the right-hand Figure small circles showing where the holes are situated when you see the target from the back.

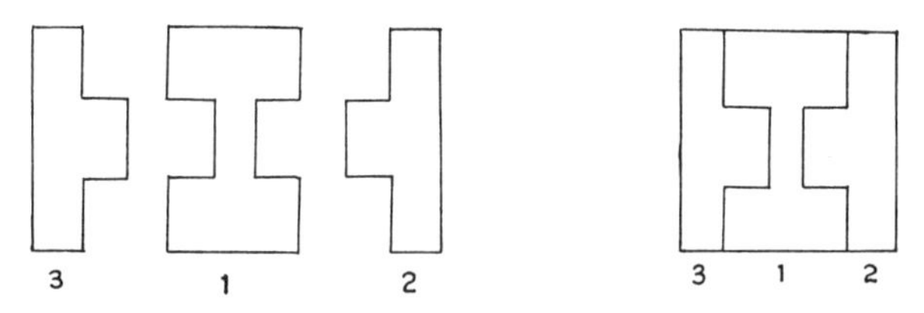

Figure 10. If you join Figure 2 and Figure 3 to Figure 1 you get a square. Now look at the figures below. If you put together two of the figures in the row under Figure 1 you will also get a square. Put a cross under the two figures which fit together with Figure 1 so that they form a square.

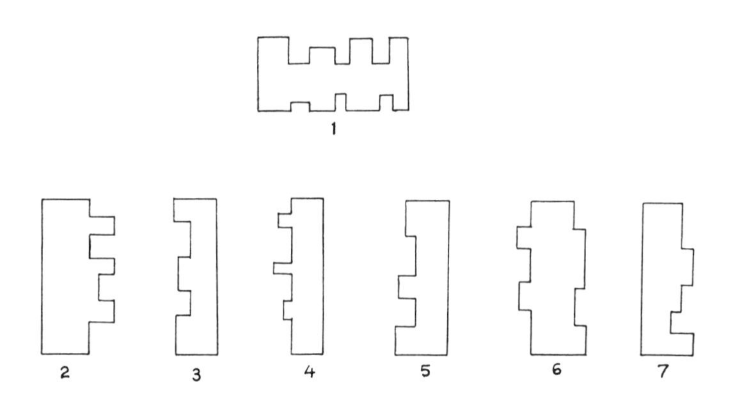

Here are three examples of verbal problems:

Why are high blocks more often built in towns than in the country?

1. Low houses are easier to build.
2. Building sites are more expensive in towns.
3. To get more sun in the top flats.
4. Diseases do not spread as easily in houses where less people live.
5. There is more space for building in the country than in the towns.
6. It is uncomfortable to live so close together.

Draw a line under the two reasons which you think are the best.

The blank spaces must be filled by numbers forming a series in which each number follows the one before according to a definite system. Now put the right numbers in the blank spaces.

6	12	18	24	
160	80	40		10
21	18		12	

In each of the following series, underline the word which does not belong with the other words in the series.

1. Pliers, hammer, nail, saw.
2. Sabre, revolver, cannon, machine-gun.
3. Train, wheel, bicycle, car.
4. Sail, bicycle, sit, ride.
5. Table, chair, sofa, picture.
6. See, hear, walk, taste.

In the history of the development of the intelligence test we can observe a move from verbal and mathematical problems to problems in pictures. Sometimes these are puzzles with two-dimensional geometrical shapes, as in Figures 10 and 11.

There are also two-dimensional forms which have to be thought of three-dimensionally, as in Figures 12 and 13.

Some tests also have a series of pictures which have to be regarded as representing three-dimensional realities, with the dimension of time added. Figure 14 is an example. In this problem you have to imagine the

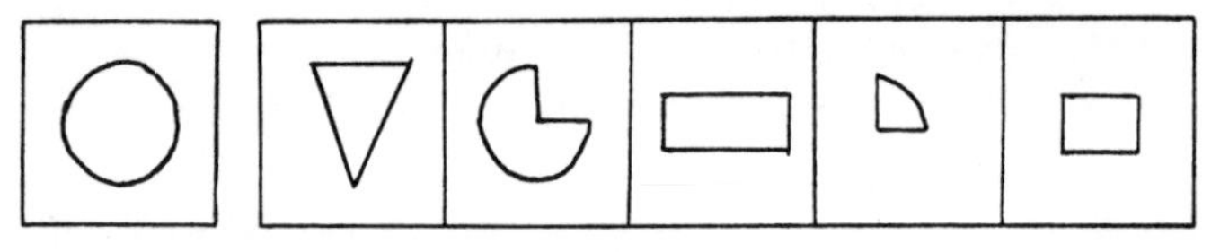

Figure 11. Put a line under the two figures in the row which together form a figure like the one on the far left.

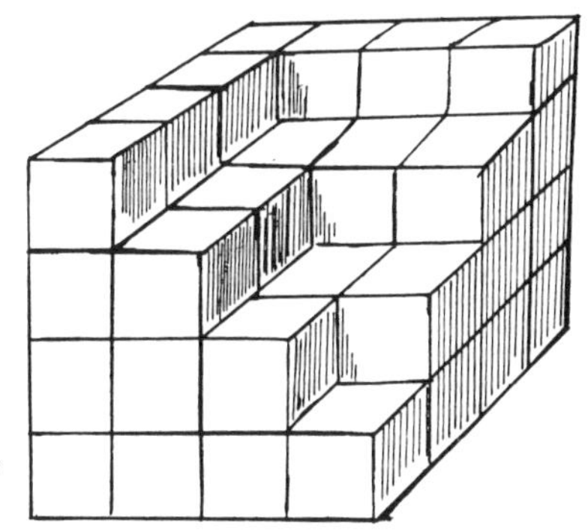

Figure 12. Block counting. How many blocks?

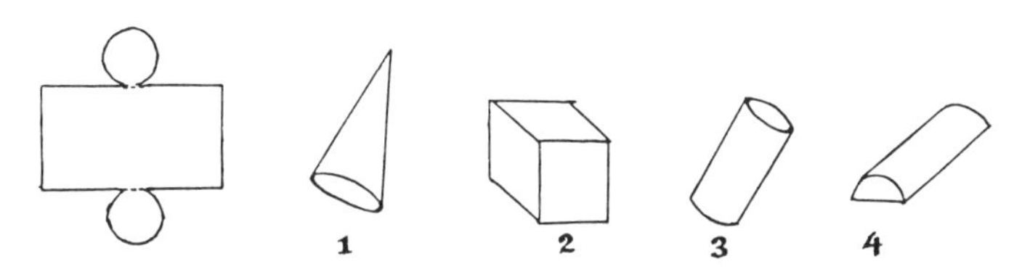

Figure 13. Sheet folding. Imagine that the left-hand figure in the row above is a thin sheet. Which of the numbered figures will you get if you roll up and bend the thin sheet?

Figure 14. Picture arrangement. Put the number 1 under the picture which should come first in the series, the number 2 under the one which should follow it, etc.

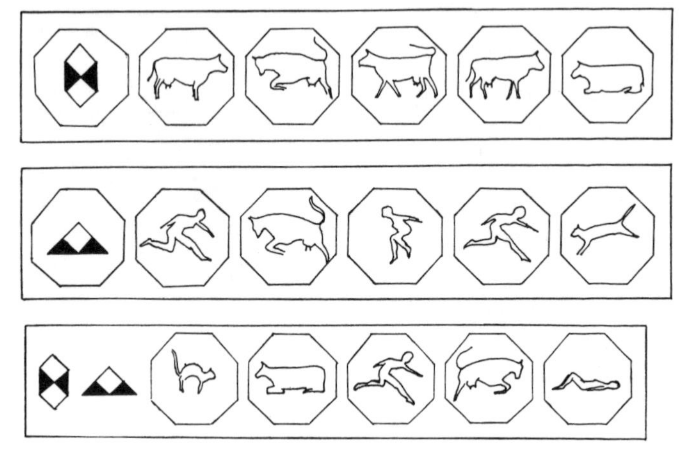

Figure 15. Problem of the type occurring in the Semantic Test of Intelligence (STI). Say which of the figures in the bottom row is symbolized by the two geometrical symbols on the left.

sequence of events from the pictures and then number the pictures in chronological order.

There was a later development in intelligence testing which looked more fruitful than the customary mental conversion of two-dimensional patterns into three or four-dimensional ones. These are tasks which test the candidate's ability to associate common features in the imagined reality represented by a series of pictures, with a symbol. Figure 15 is an example of such a problem.

This type of problem was designed at Harvard University at the request of the American army. The army wanted to produce a test which could be used to predict which of their illiterate recruits would achieve the best results from being taught to read and write. The resultant test, "The Semantic Test of Intelligence", proved better able to predict scholastic success than "Army Beta" (the First World War Army group intelligence test for illiterates). Scholastic success was measured by the teachers' evaluation of the progress of the illiterate recruits.

The problem in the example from the STI in Figure 15 can be solved in two ways, one more concrete and one more abstract. If you spend some time looking at the pictures in Figure 15 you will be able to say which picture in the bottom series represents a situation of the type denoted by the symbols. Do this before reading on.

You may have solved the problem in this way: you have seen that the picture sequence following the hexagonal symbol and the picture sequence following the triangular symbol have one picture in common: that of the jumping cow. The jumping cow is the only figure found in all three pictorial sequences. So you selected the picture of the jumping cow.

If you solved the problem in this way you may not at once be able to solve the problem in Figure 16.

Or you may have solved the problem in Figure 15 in another way, in which case you will need no extra time to think about the problems in Figure 16. You will have seen that in Figure 15 all the pictures in the sequence following the hexagonal symbol represent a cow and that all the pictures following the triangular symbol represent a living creature which is jumping. So you have given the description "jumping cow" to the sequence following the hexagonal and triangular symbols in the bottom row. Then you will at once understand the point of the problem in

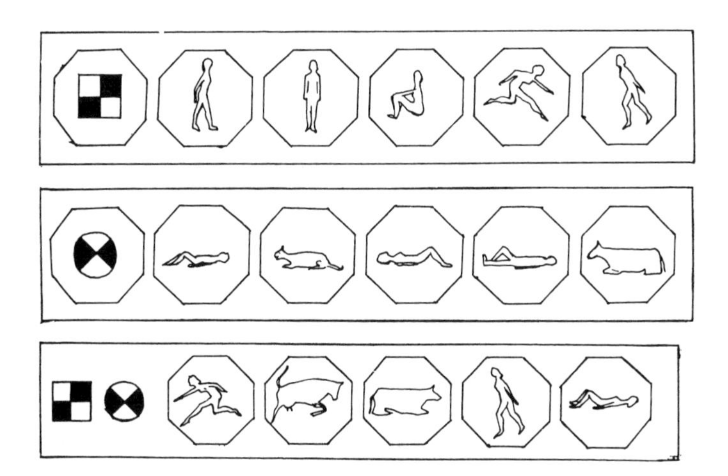

Figure 16. Problem similar to the one in STI. Say which of the figures in the bottom row is symbolized by the square and the circle on the far left.

Figure 16. But the STI test is unique. To the best of my knowledge, there has been no development of an intelligence test on the lines of the STI since 1952. Instead, modern intelligence tests have been designed to measure the different intelligence factors. These factors are described in more detail in the section on factor analysis, Intelligence and Intelligence Factors, in the Appendix at the end of the book. For the present we can say that the most usual factors are a logical factor, two verbal factors, an arithmetical factor and a factor which determines how clearly you understand what you see.

What is perhaps the most modern of Swedish group intelligence tests was designed at the Pedagogic Institute of Lund University. The test, known as WIT III, is timed (45 minutes) and contains five types of problem. The first type of problem is primarily intended to measure the logical factor. There are sets of five figures in a row, all but one of which have quantifiable features in which they resemble each other. That is, the figures resemble each other in their geometrical properties. Figure 17 is an example of a problem of this type. Four of the geometrical figures resemble each other in that precisely half the surface of each figure is shaded. The fifth is different.

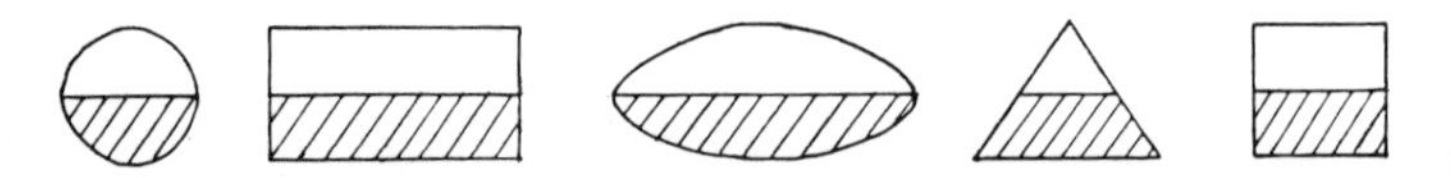

Figure 17. Classification. Put a cross under the figure in the row which does not belong with the others.

The next type of problem consists of word analogies. A word-pair is given. The candidate chooses from among five words a word-pair analogous to the given word pair. Example:

Stable – stall guard, lock, cell, prisoner, prison

This problem is intended to measure the candidate's command of one verbal factor.

The third type of problem consists of contrasts. From a sequence of five words the testee must select two words which mean the opposite of each other. Example:

Longer, less, higher, more, worse.

This problem is supposed to measure the second verbal factor.

The fourth type of problem is supposed to measure the arithmetical factor, known in the specialist literature as the numerical factor. From a number of figures (from three to five) the testee has to produce a total in which one of the figures is the answer and the rest make up that figure. Any method of calculation can be used. Example:

2 3 5 30

The last type of problem is designed to measure the perceptual factor. The problems consist of a jigsaw puzzle. Figure 18 gives an example.

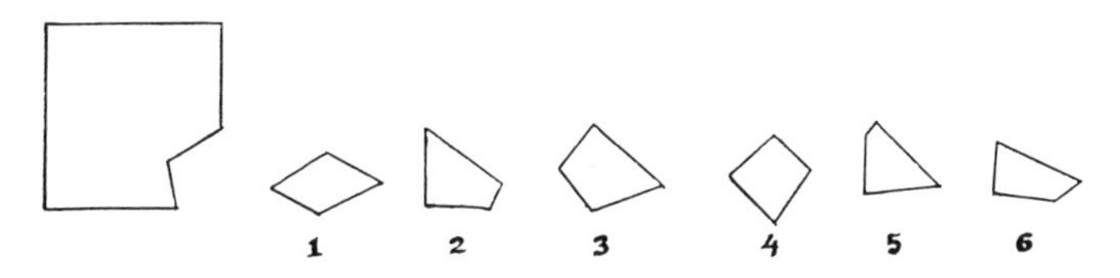

Figure 18. Jigsaw puzzle. Put a cross in the numbered figure which fits together with the large figure so that it becomes a square.

The group tests from which examples have been taken are conventional in the sense that the problems are of a type which occurs in most intelligence tests. There are also less conventional tests. One is Goodenough's Draw-a-Man Test, designed in 1926. The testee was asked to draw the best picture he could of a man. After a revision in 1961 the tasks of drawing a picture of a woman and a picture of oneself was added. The testee gained points according to a definite system for all the details included: eyes, right number of fingers, clothes, etc. Perspective, and the position of the body in relation to the assumed ground, were also evaluated. The correlations between the results of the Goodenough and Stanford-Binet tests are low –

Figure 19. Goodenough Draw-a-Man Test. Human figure drawn by 6 year-old girl. Mental age 6. IQ 100.

between 0.40 and 0.80, depending on the heterogeneity of the groups studied. Figures 19 and 20 show the test results for two little girls.

What made the task of drawing a picture of a person into an intelligence test was a standardized rating applied to a large group of children and grown-ups of all ages. Goodenough studied how Western children of a particular age draw a person, and he knew how mentally defective children draw. He believed for a long time that the test could be used to compare the inborn mental capacity of people of different cultures.

The Porteus Maze Test, which was designed in 1924, is

Figure 20. Goodenough Draw-a-Man Test. Human being drawn by girl with a real age of 5 years and 9 months. IQ 120.

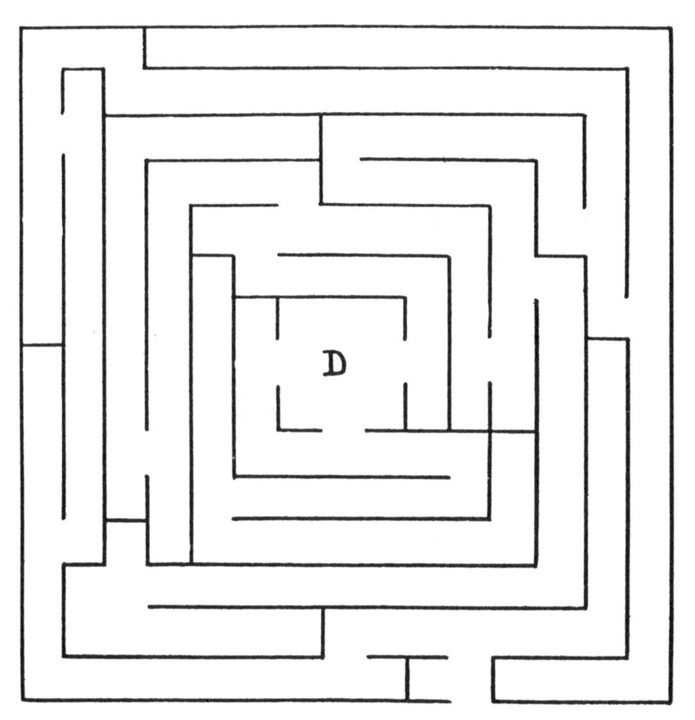

Figure 21. Porteus Maze Test. Problem for adults. Draw a line from the "entrance" to the centre of the maze without lifting your pencil from the paper and without crossing any of the lines.

different. It consists quite simply of pages of more or less complex mazes, marked out in lines. The problem consists in drawing a line from the entrance to the goal without lifting the pencil from the paper and without crossing any of the printed lines. Porteus thought his test measured the capacity to plan and foresee. Like the Goodenough test, it is untimed. See Figure 21.

The values used in England to interpret data in connection with factor analysis of intelligence-test results differ from those used in the USA. Many English psychologists consider that intelligence consists of a general intelligence factor, the *g*-factor. In 1938 Professor Raven began to develop his Progressive Matrices Tests. These are non-verbal group tests, consisting of patterns from which a

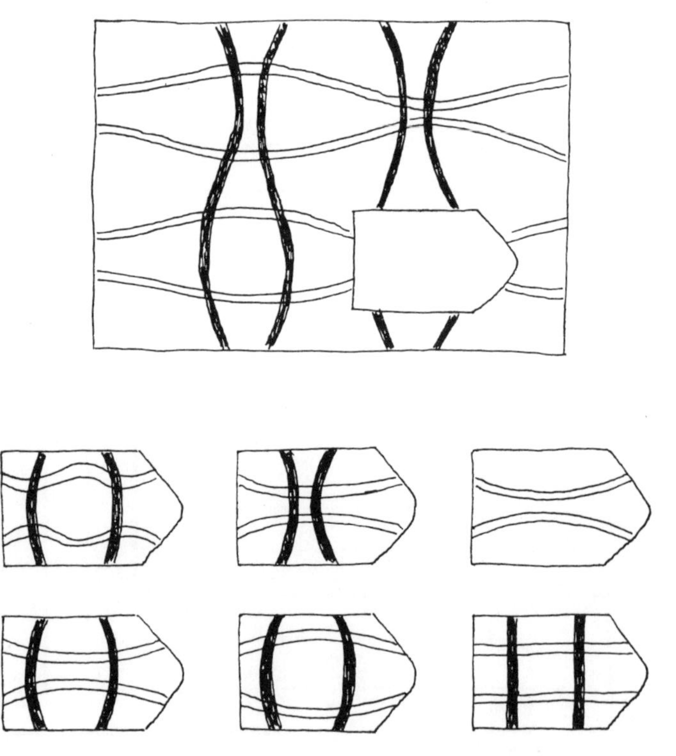

Figure 22. Professor Raven's Progressive Matrices Test consists of problems of this type. The testee has to say which of the six portions under the large pattern fits into it.

piece is assumed to have been cut out. Pieces of various patterns, of the same shape as the section cut out, are shown to the testee, who has to select the piece which fits the pattern. Figures 22 and 23 give examples of problems from Raven's Matrices.

Professor Raven claims that these tests measure inborn mental capacity better than Stanford-Binet. The correlation between the two tests is about 0.86, according to Raven, but considerably lower according to Professor Anastasi. The test is not timed.

Davis-Eells Games is the name of another type of intelligence test, originally intended to measure the intelligence of children from lower social groups. The test consists of pictures similar to those in newspaper cartoons. The idea was that the child's verbal handicap should not

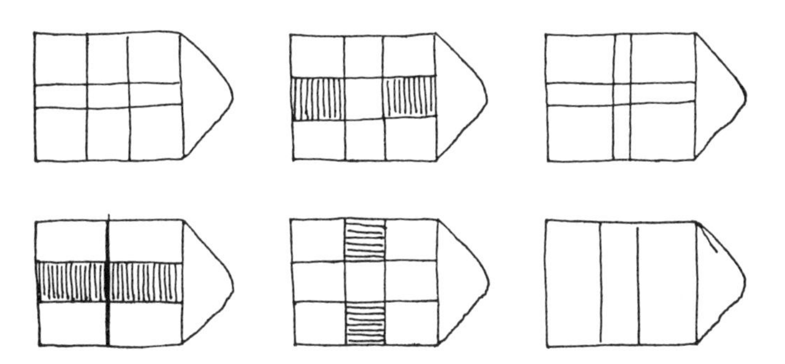

Figure 23. Problem from Raven's Matrices.

Figure 24. Problem from Davis-Eells Games. Which of the girls is putting the bottles into the black crate so that the white crate will stand most steadily when it is put on top of the black one?

influence his IQ score in the Davis-Eells Games. A handicap did still exist, however. The correlation with Stanford-Binet is about 0.50. Examples in Figures 24 and 25.

Thurstone's Colour-Word Test is a relatively new intelligence test. The testee's capacity to resist the *suggestions* involved in *reading a word* is measured. First he is given a list of names of colours to read through: "Blue, green, yellow, red, black, blue, white," etc. In the next section of the test he is given a list with borders of different colours, which he has to name. Finally we come to the test itself. It consists of a list similar to the first. The difference is that the name of the colour is printed in another colour, different from the one named. The word blue is printed in green, the word green in red, etc. The problem is to write down the colour of each word. The test is, of course, timed.

Group tests are designed with special instructions for their administration, so that they can be given to anyone. Even non-psychologists must be able to give the test to a large group of people at the same time. This can be difficult.

It is often important that exactly the same words and emphasis be used in giving the test instructions to all the testees, if their results are to be fully comparable. For instance, you might be asked to draw, on a piece of paper representing a field, the best route to take through it in order to find a lost ball within an area (a problem from the Stanford-Binet test). You gain marks only if you draw a

Figure 25. Problem from Davis-Eells Games. Put a cross before the sentence which you think fits the picture best.
1. The boys want to clean the man's window and pavement.
2. The man is forcing the boys to clean the window and the pavement.
3. It isn't possible to say why the boys are cleaning the window and the pavement simply by looking at the picture.

To give the right answer calls for familiarity with pictures. Not everyone will see at once that one of the boys is washing off a cowboy's head on the pavement and the other an old man drawn on the window.

path which shows that you are using a definite system in your search. If the examiner says, "Show what system you are using to find the ball," he is helping the testee by suggesting a systematic process. Figure 26 shows an acceptable and an unacceptable solution.

Another example: a boy with a Stanford-Binet IQ above 130 took a group test in which his IQ was put at 108. It

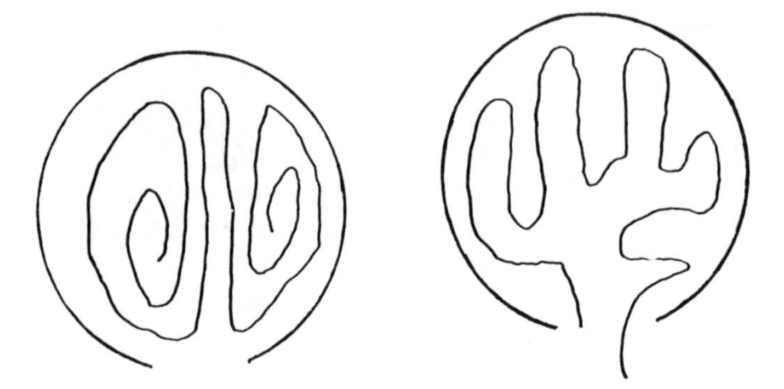

Figure 26. Problem for 12 year-olds in Stanford-Binet. "Let's suppose that your purse with a lot of money in it has been lost in this big field. Take this pencil and start at the gate and show me where you would go to hunt for the purse so as to be sure not to miss it." According to the scoring standards, the solution on the left is satisfactory but the one on the right scores a minus.

turned out that the teacher who had given the test had warned the boy to make absolutely sure that the little square on which the correct answer was to be marked was completely filled. The boy, who was very conscientious by nature, had taken a great deal of trouble to fill in the small squares so that they were completely black. He even went back again to check that the squares were correctly filled. Since the test was timed, not many problems were solved and his IQ rating was 108. Now, we could say that the boy with the IQ rating of 130 from the Stanford-Binet test was not particularly bright after all, since he could not see for himself what was essential in the test, but whatever the reason, it indicates that differences in test results sometimes reflect differences in the examiner's behaviour and attitudes.

Individual tests are usually given by specially trained psychologists. They take longer to complete and the results seem more reliable. That they really are so is not always true, because the marks given in individual tests depend more on the examiner's judgment and evaluation than the marking of the group test.

A study of different marks given by different examiners (the Harvard Growth Study) showed that one of the examiners studied gave IQ scores 13 points higher *on average* than the examiner who was meanest in his marking. In individual cases differences of from 30 to 40 IQ points were reported. In the section on the IQ of different population groups we show that white examiners have a strong tendency to give white children higher IQs than Negro children and the same applies conversely to Negro examiners.

The American psychologist Rosenthal carried out the following experiment, which aroused great interest. A group of examiners were told to test the responses of trial subjects to photographs of human beings. Half the examiners were told that the subjects would probably turn out to respond positively to the photographs. The other half were told that there was reason to expect that the subjects would respond negatively. The examiners were not told that the subjects had previously been tested with the same photographs and had been completely neutral towards them. The results of Rosenthal's experiment were that the first group of trial subjects were judged by the examiners to have positive attitudes to the photographs. The subjects tested by the other examiners were reported to have had negative attitudes. The difference was so great that there was no demonstrable overlap between the first and second groups. That is, no one in the negative group had an attitude as positive as the least positive in the positive group.

In another experiment Rosenthal showed that rats were judged to be more or less intelligent in the maze test, depending on the expectations the conductors of the experiment had been given before the beginning of the test. Half of the experimenters had been told that the rats in one cage belonged to the most intelligent strain. The other half had been told the opposite.

Expectations of this kind affect the examiner's evaluation of the answer in verbal tests more than others. If the question is "What is meant by: Strike while the iron is hot?" (WAIS) and someone from a low social group answers "Iron your clothes before the iron cools down," he is unlikely to score any points. But a University student will probably get full marks for this answer: "Do whatever it is you're going to do quickly, before the favourable conditions disappear." Both are equally well able to see that the saying means that you should do things quickly, before the favourable conditions for doing them have disappeared, but because they express it in a more or less sophisticated way they get different marks.

The most important individual tests are the Stanford-Binet and the WAIS and WISC (Wechsler Adult Intelligence Scale and Wechsler Intelligence Scale for Children).

Another important distinction between different types of tests is the distinction between verbal types and performance types. A typical performance test problem is Seguin's Form Board, consisting of a board in which holes of different shapes have been cut, with pieces of wood which fit the holes, either singly or with other pieces. Another performance problem is the Healy Picture Completion Test, consisting of pictures from which a piece has been cut out, such as a large picture of a crowd of children playing outside a house, with trees and animals and so on. Square holes have been cut in the picture, which is supplied together with small square pictures of different objects which have to be fitted into the holes. In Kohs Block Design Test the testee receives a number of cubes of the same size, with their sides painted in different colours. These cubes must then be arranged in the form of a pattern which the testee is shown on a piece of paper.

Figure 27. Seguin's Form Board occurs in a number of intelligence tests for children. It consists of a board with holes cut in it and blocks which fit the holes.

There is a definite time limit for each pattern and additional marks are awarded to the testee who completes the pattern first.

In Stanford-Binet, WAIS and WISC, performance problems are used for small children. Stanford-Binet has no performance problems for the older age group, but WAIS is divided into a verbal section and a performance section.

6

DO INTELLIGENCE TESTS REALLY MEASURE INTELLIGENCE?

Validity criteria are the comparative objects used to decide whether a problem really measures what you want it to measure. Intelligence tests are intended to measure intelligence. In order to be sure that a problem, e.g. "What does the word 'jealous' mean?" (Stanford-Binet), measures intelligence, we must know whether people of the age for which the test is designed can define the word jealous in the way the designer of the test regards as correct, without their parents' vocabulary or their schooling or other factors apart from "real" intelligence playing a major role.

This means that although the psychologists have no precise common definition of intelligence, they do have a common idea of the kind of thing intelligence is. When they select problems to measure intelligence they are acting on this common approximation to what intelligence is.

In practice this means that the designers of intelligence tests collect problems similar to those which occur in existing tests. Such problems are said to have face validity. But sometimes someone designs a problem of a type which has never occurred before. For instance, Goodenough devised an entire intelligence test on the problem "Draw a picture of a person," and many problems in modern tests were once quite novel types. Where this kind of problem is concerned we consider that a high correlation between the results of the new problems and the results of well-known intelligence tests guarantees that they are problems requiring intelligence.

In order to be fully satisfactory the correlation must be above 0.80. The correlation between Stanford-Binet and WAIS, the two best known individual tests in the USA, is

0.85 – 0.95 in large, heterogeneous groups. However, most intelligence tests have a much lower correlation with each other, somewhere between 0.50 and 0.85. One has only to glance at Figure 5 to see the extent to which the tests measure the same functions.

It is not clearly understood that an intelligence test is not intended to measure an aptitude for passing intelligence tests. In that case no criteria of validity would have been needed. The question: "What validity has this test?' would be answered quite simply by studying the problems in the test.

The test is intended to measure another attribute – intelligence. Opinions differ as to how important it is to have an exact definition of intelligence. An American Professor of Psychology has said that this is why many psychologists have now reached the point where they no longer ask themselves: what is intelligence? They have decided that they can do useful work by measuring intelligence without defining it. In this respect they are doing the same as earlier physicists did when they were studying heat. Long before the physicists had agreed on a reasonable definition of heat they had invented reliable thermometers with which they could measure changes in temperature and with these instruments they were in a position to discover many important physical laws.

Another American Professor of Psychology writes: "If we do not define 'intelligence' . . . , we can never prove or disprove that a particular test is a valid measure of 'intelligence'."

As we have already indicated, intelligence researchers *have* a common idea of what the test results they want to produce should be like. And in broad outline this is a definition of intelligence.

> *Firstly* the test results from a large group of people, representative of a whole population, must be divided up in the same way as body measurements. (The group must not contain proportionally more of any particular group than the population as a whole.) This means – if IQ units are used (other kinds of markings are generally used now but the principle is

the same) – that about 68% of the group must have IQ ratings between 85 and 115. About 13% must have scores between 70 and 85, the same number between 115 and 130 marks and 2–3% will have an IQ score of less than 70 or more than 130. (See "Intelligence and the Normal Curve". The normal distribution is described there in more detail.)

Secondly: there must be no difference between the average test results of men and women of the same age.

Thirdly the IQ must not show any increase with chronological age. But it must decrease slightly after the age of 20 to between 50 and 60, and thereafter rather more sharply.

Fourthly, when compared, the average test results must reveal clear differences between different social groups. The lower social and economic groups must have lower marks than the higher ones.

Fifthly, the test results must show a clear and close relationship with school results and success in University studies and professional work.

The first four requirements as to the distribution of test results and their association with sex, age and social group are more or less technical. But the fifth requirement assumes that the problems are of a quite distinct type. The problems must have a straightforward connection with what is taught at school and University. And so they have. Professors Thorndike and Hagen: "The correlation between school education and intelligence test results is high. In view of all the hundreds of correlation coefficients reported, a figure of 0.50 to 0.60 can be taken as fairly representative. . . . Higher correlations have been encountered at the primary stage than in secondary school and University. Previous studies have shown a drop in the correlation from 0.70 in the primary stage to 0.60 in secondary school and 0.50 in University. The drop in the correlation coefficients is probably a result of the more limited difference in intelligence between the individuals in the University groups."

The correlations between IQs and school marks are

generally highest for marks in the mother tongue. The correlation coefficients for the Otis Intelligence Test and school marks for a group of American pupils, according to a very representative study, were: English 0.74, Algebra 0.68, General Science 0.64. The other subjects had moderately high correlations. The lowest correlations were in Chemistry, Latin and German, at about 0.30. The problems in the Otis Test are principally concerned with words and meanings: verbal problems.

On the relationship between intelligence and school success, an American special study on tests commented in 1964 that academic success was evidence of mental ability, although it might also be affected by factors other than intellectual ability. On the whole, the more competent people completed a higher formal education, reaching higher levels at school and at University.

This common idea about the distribution of the results of intelligence tests and their relationship with other factors has arisen from the view of humanity held by Galton and Binet and their contemporaries. It has developed and taken on sharper outlines, as data from the existing tests has poured in. The existing tests have acted as yardsticks for the idea of what intelligence is. I quote from the same special study as above:

"As individuals age their intelligence level increases, until the adult maximum is reached. Really well known groups, such as the gifted, the somewhat superior, the average, the slow learners and the mentally defective must clearly show different levels of performance in order for a test to be regarded as having high validity. A new test must correlate well with another instrument of proven validity designed for the same purpose."

"Another instrument of proven validity" generally means the Stanford-Binet test.

7

HOW RELIABLE ARE INTELLIGENCE TESTS?

To question a test's validity inevitably also means questioning its reliability. If there is to be any significance in calculating validity coefficients and other correlations, the test results must be very reliable. A boy who gets an IQ score of 100 in a test on one day must get almost the same result if he takes the same test a few days later.

Intelligence test results must be reliable. This is essential. At the same time, this assumes that intelligence is an attribute which does not fluctuate from day to day in the same way as aggression, or determination, or the ability to solve chess problems.

Reliability, like validity, is measured with the help of correlation statistics. If the results obtained by one group of people agree with the results they obtain with the same test a few days later, so that the correlation is over 0.90, the test's reliability is regarded as assured. A test with a correlation figure of 0.92 is regarded as extremely reliable. From Figure 1, which shows the reliability of Stanford-Binet-37 for 7 year-olds, we know that this means that a girl who got an IQ score of 105 one day may get anything between 90 and 115 for the same test the next day. The reliability is greater for lower IQ values. A boy with an IQ score of 70 on one day is unlikely to get more than 75 or less than 65 when tested again.

In general, the reliability coefficients for Stanford-Binet, WAIS and WISC are just over 0.90 on immediate re-testing, or where two parallel tests are taken one after the other by the same individual. The reliability of most other intelligence tests is considerably lower: 0.70 and 0.80 are representative figures.

For large groups of people intelligence test results remain more or less as constant after a few years as after a few days. The psychologist P.E. Vernon writes: "The correlation between similar (not necessarily identical) tests over the 6 to 11 year-period, or over 11 years to young adulthood, does not normally drop below 0.70." This means that half of all the people tested will have more than 7 points more, or more than 7 points less when re-tested after 5 years. The intelligence of 17 out of every 100 tested will be more than 15 IQ points higher or lower, and that of one person in 100 will be more than 30 IQ points higher or lower.

It is generally true that the lower the intelligence, the more stable it is. This is probably largely the result of the community's attitude to someone who has once been measured and found to have below-average intelligence. He is rarely given a chance to change his status. Think of the marking systems in schools which effectively take the stuffing out of the pupil with a poor intellectual background right from the start. The constant competition to which this system of marks gives rise singles him out from the rest, until he loses the little interest he may once have had in education, having been left at the bottom of the class the whole time, constantly compared with the others in the class, so that his intellectual development stagnates. Another example of the community's attitudes in this connection is displayed in one of the most popular introductory psychology manuals at Swedish Universities. Professor Norman L. Munn writes: "Infants who can be presumed to be capable of developing into bright children should be placed with the most intelligent adoptive parents. The duller children should be placed with adoptive parents who are not particularly intelligent and who will consequently not be too disappointed at what the child can achieve." (*Psychology*, 4th impression, 1961.) In a later section studies are reported which show that the adopted child's IQ rating is closer to that of the adoptive parents than that of the biological parents in cases where the difference between the social milieu of the biological mother and the adoptive mother is considerable. This means that if infants are placed with intelligent adoptive parents they will become intelligent.

It should also be noted that the older the testee, the

more stable the IQ. In other words, the older the examinee at the first test, the more likely it is that he will get roughly the same score in the next test. Tests of small children under 3 are virtually completely unreliable. The values derived from tests with 4 and 5 year-olds say very little about their test values when adult. Even up to the age of 10 it is true to say that the most reliable predictions of a person's IQ at 25 will come from noting the parents' profession and social group and their plans for the child's future education.

8

INTELLIGENCE TESTS AND VALUE JUDGMENTS

From a strictly logical standpoint, any line drawn between knowledge and values is arbitrary. A good example of this is a mental examination in which one human being sits in judgment of another's perception of reality.

Let us assume that two men are standing in a field. "There is an oak tree," says one, pointing. "No, there's no oak tree there," says the other. There are several ways of deciding whose perception is true. The disagreement may cause a fight in which the one who kills the other believes this proves that his perception of reality was the right one. Or a third party may be called in, so that the majority decision determines which reality is the "true" one.

The closest we human beings can come to the "true" reality is to see it, hear it, touch it with our skin. It makes no difference if we possess the world's most effective machine for determining the presence of oak trees. After all, it is the individual's sensory impressions from the machine's measuring counter which in the last analysis decide whether the oak is there or not. Every perception of a circumstance as factual, or true, involves a valuation of one's own or someone else's sensory impressions as correct or incorrect reflections of reality. Fundamentally every valuation is based on (presumed?) knowledge.

The fact that I value a bird in the hand more than two in the bush means that I know that I am not going to be able to catch one or more of the birds in the bush.

By "value judgments" I mean that things, circumstances in life or symbols, are given a significance which is not obvious to other people. In this use of the word "value", all symbols involve values, apart from those whose significance is directly understood by the other person or people.

The fact that intelligence tests and value judgments are associated may, in itself, be uninteresting. What is most interesting is *what values* are expressed in the test.

Underlying all intelligence tests are two value judgments, which can be expressed like this:

1. Many people can think better than others.
2. It is important to discover who think worst and who think best. (This is the basic value judgment.)

Now for the values which emerge from individual test problems. In Stanford-Binet there is the following problem, intended for superior adults: "Give three reasons why a man who commits a serious crime should be punished." In the comment under the problem we read: "It occasionally happens that the subject says he does not believe a man who commits a serious crime should be punished. In such a case, ask him to give the commonly accepted reasons."

According to the scoring standards, acceptable answers will include: "To make him suffer for his wrong-doing," "To teach him a lesson."

This problem is based on a fundamental value, or principle, which is regarded as self-evident: someone who has committed a serious crime should be punished. An intelligent person who does not share this view may fail the test because by independent thought he has arrived at reasons why a serious criminal should not be punished which are not "the commonly accepted reasons." The idea that the criminal should be punished because punishment involves suffering, and because the criminal must suffer, are obviously unreasonable to a modern man. The only things in which the community is really interested (apart from the virtuous citizen's possibly perverse desire to know that someone else is sufferıng) are that as few new crimes shall be committed as possible and that the steps taken to achieve this shall cost as little trouble and labour as possible. In other cultures these problems have been solved by, for instance, cutting off thieves' hands. This was done not only to make them suffer but also to prevent recidivism. The idea that a thief should suffer, arises naturally from Christianity and the story of Jesus' suffering. The philosophy of it is that *by physical and mental suffering* on a cross-shaped instrument of torture on Golgotha, Jesus was *atoning for crimes* committed by mankind. This problem is an example of a value judgment which assesses Buddhists and Hindus as people of inferior intelligence to those from Christian cultures.

Another fundamental attitude can also be sensed here, which may be even older than the first: return evil for evil.

Every verbal problem tests the subject's own values. There is no word which has an established and given meaning, once and for all. Language is as alive as the people who use it. It is always changing. If there had been intelligence tests in the 1700's and if the question "What is a pot?" had been asked, a modern person would have been found less intelligent. "Pot" at that time meant a clay, china or glass vessel in the most general sense, a vase or flower-pot – not a toilet article. The norms which apply to the word definitions in Stanford-Binet and WAIS, i.e. what you must say in order to pass, reflect a certain way of using the word in a certain stratum of society at a certain period of time. Consequently individuals from this stratum of society and individuals from more conservative linguistic environments automatically come out a little more intelligent than others. The word definitions which pass arise from the scoring standards for the test. These standards are then valid for a twenty-year period, during which the language alters slightly. Those who have a keen ear for changes in language will get lower marks in the word definition test than those from linguistically conservative environments.

For individuals these factors fade into insignificance, by comparison with the quality of schooling and parental attitudes. But for correlation coefficients describing conditions for large groups these factors are essential. For instance, the fact that, on average, lawyers have higher IQs than artists in the American data may result from factors such as these.

The constant change in language is significant not only in the word definition tasks but to all verbal tasks. Because of this, more recent intelligence tests attempt to put verbal tasks aside and use figures, lines and points which stand in a numerical and geometrical relationship to one another. The section on "culture-free" tests will show that not even this type of problem is free from conventions, and that people who have grown up in cultures dominated by the alphabet and Euclidean geometry will, according to these tests, be more intelligent than others.

Obviously, word definition tasks primarily test knowledge, not values. The Stanford-Binet question "What is a layman?" should not be answered by "Someone who looks after hens," although this may sound right to someone who is unfamiliar with the meaning of the word. The correct answer would be: "Someone who is not a professional expert", or: "Someone who is not a priest".

The simpler and more obvious types of value judgment expressed in verbal intelligence tests are easier to study if one is oneself in an environment other than that to which the test belongs. I am convinced that a Chinese, for instance, can recognize the values implicit in test problems more clearly than a European. As Marshall McLuhan has said, "Environments are invisible. Their basic values, their inherent structure and general patterns are impossible to grasp." I shall have to restrict myself to those values which stand out as such to me, born as I was in 1938.

Figure 28 shows a problem from a group intelligence test for primary school children in the USA. It shows our Western civilization's value judgment regarding rectangular and symmetrical patterns.

The following problem from Stanford-Binet is intended for 13 year-olds: "A man who was walking in the woods near a city stopped suddenly, very much frightened, and then ran to the nearest policeman, saying that he had just seen hanging from the limb of a tree a . . . what?"

A child who replied "A monkey" would get no marks. According to the instructions for the assessment of answers: "The expected response is 'a hanged man.' " The

Figure 28. Put a cross under the nicest house. Problem from the Pintner-Cunningham Primary Test, a group intelligence test for children.

example shows the high evaluation of distrustfulness. It is more intelligent to believe the worst than to believe the best.

One of the questions in the section on "comprehension" in the WAIS test is: "Why should we avoid bad company?" The answer "It leads to temptation" gets two points, while the answers "It leads to trouble" and "You might get involved in a situation you couldn't control" fail the test.

Intelligence tests generally offer far too many examples of answers which are quite sensible, indicate clarity of thought and realism and are moral from the general human standpoint but which fail because they do not correspond to some religious or philosophical standard. The Stanford-Binet test contains this problem for ten year-olds: the child is shown a picture of a white man aiming his gun at an Indian standing some way off, while two Indians closer to him are about to attack him. The question is: "What's foolish about this picture?" The testee who gives the answer, "The soldier is shooting at the Indian coming towards him without noticing the ones coming at him from the side," is intelligent. But the following answer is regarded as a sign of inferior intelligence: "The clothes the man in the picture is wearing are too modern for the time of the Indian wars; his gun is too modern."

The following problem is set for eight year-olds in Stanford-Binet: "What would you say if you were in a strange town and someone asked you the way?" The little boy who says, "I would say I didn't know the address and I would ask Mummy where it was," gets full marks. But this is not a particularly bright answer, because if the town is strange to the boy it is probably strange to his mother as well. A Swedish child with a capacity for independent thought might answer, "I would give him a telephone directory." (There are maps in Swedish telephone directories.) This little boy's answer is unintelligent according to the evaluation standards of the test.

Nine year-olds are given the problem "Give two reasons why children should not be too noisy in school." A correct answer is: "Because they would be punished. They would be made to stand in the corner." No points are given for: "If they made a noise in school it would be heard everywhere and disturb the other classes."

WAIS has the question "What should you do if you are at the cinema and you are the first to notice that a fire has started?" Two points (maximum marks) are given to the reply "Tell the cinema manager." The answer "Go to the nearest exit" gets no points, which is understandable. But the answer "Go and fetch some water" also gets no points. Couldn't he have received one point for such a positive idea? In any case, is it really all that bright to start looking for the cinema manager in this situation?

The assessments of answers are sometimes so odd as to be almost incomprehensible. The Stanford-Binet test for eight year-olds contains the following question: "About two o'clock one afternoon several boys and girls in their best clothes rang the bell at Alice's house. Alice opened the door. What was happening?" Among the unacceptable answers were: "They wanted to play with her" and "They wanted her to go somewhere with them." These answers passed: "Some boys and girls had come to visit her," "They were going to do a play," and "They wanted to give her a surprise."

On the whole, the test designer wants the answer in a quite specific form. Consistent, well-formulated answers which do not correspond to this preconceived form are generally unacceptable. Symmetry often plays a major role: in the problem "A bird flies, a fish . . . ?" the answer "lives in the water and swims" is not allowed. The problem comes from the Stanford-Binet test for five year-olds. In a similar problem for above-average adults: "Pines are evergreen, poplars are . . . ?", "deciduous" gets a mark, but "deciduous trees" fails. In a test for eight year-olds the question is: "Wolves are wild, dogs are . . . ?" The expected answer is a word meaning the opposite of "wild," but the words "gentle," "playful," "friendly," "quiet" get no marks.

In other cases the assessments are still more inexplicable. In a problem for fourteen year-olds in Stanford-Binet the testee is shown a picture of the sun, a human being and his shadow. Question: "What's foolish about this picture?" Answer: "The shadow, because it's facing a different way from the person" is not accepted. "The shadow is going the wrong way" gets a pass. Averagely intelligent adults are asked: "What have 'tall' and 'short' in common?" The answer: "Both are descriptions" gets no marks, whereas "Both are measurements" passes.

"What does milksop mean?" is a vocabulary test included in the range for many age groups. The answer "It's an expression of disgust" gets a minus. "A slang term given to a rich person or a person that is used to having other people do things for him" is a response which is regarded as a sign of intelligence.

This is just a sample of the frequently incomprehensible values underlying the norms for assessing answers in Stanford-Binet. This test has been used in psychological studies and as a selection tool and has affected the lives of tens of thousands of people.

The evaluations of answers in the verbal section of WAIS contain as many oddities as Stanford-Binet. In answer to the question: "What have an orange and a banana in common?," "Both are fruit" gets twice as many marks as "Both have peel." The answer to "What have an egg and a seed in common?" should be "They are the origin of life," but not "Something grows from both of them."

9

NO INTELLIGENCE WITHOUT KNOWLEDGE

Galton's ideas about the inheritance of mental capacity from ancestors means that a human being who has been unlucky enough to grow up without either books or contact with other human beings would still have an intelligence level which was determined at conception. Unfortunately this could not be tested. All intelligence tests assume some knowledge.

There are eleven sub-tests in WAIS. Of these, the memory test, the mixed symbol test and Kohs Block Design Test (described in the next chapter but one) are completely independent of knowledge of Western civilization. The highest score obtainable in the WAIS test is 437. Only 35% of this total comes from the completely knowledge-free tests, and more than half of the total marks derive from timed tests. In order to be intelligent according to WAIS, you need both knowledge and speed.

The Stanford-Binet test is not timed and does not seem to require knowledge to the same degree as WAIS. Almost half of all the problems are concerned primarily with memory, the ability to draw logical conclusions and the ability to detect absurdities. The rest cover the ability to manipulate and define words and meanings. But I think it would be wrong to assume that Stanford-Binet requires proportionately less knowledge than WAIS. Even a problem such as "A rabbit is timid, a lion is . . .?" requires for its correct solution a certain family background and probably attendance at a kindergarten. How is someone who has grown up in the slums to know that a lion is brave? It might be more likely to occur to him that a lion was strong or savage. And a problem such as: "Why is this foolish: A soldier complained that everyone in the regiment except himself was marching out of step" is an impossible question for a nine year-old who thinks that "marching in step" means "putting his foot down when the drum beats" or "making the abrupt movements soldiers usually use." You have to know *exactly* what marching in step means in order to get any marks for your answer. It is probably no exaggeration to say that more than half the problems in Stanford-Binet call for special knowledge, quite apart from knowledge of Western logic, for their solutions.

This means that children from lower social groups automatically get lower scores, regardless of their actual ability to make use of Western logic in their thinking.

Raven's matrices is one of the very few tests which does not assume any knowledge other than an awareness of Western logic and its basic values (the high valuation of the symmetrical relationship). Nor is this test timed. But this means that to a great extent is has come to measure perseverance, interest in the problems and intellectual self-confidence. More about this in a later chapter.

The correlation between the results (IQ scores) from Raven's Matrices and the marks from WAIS and Stanford-Binet is roughly 0.65, which means that most testees achieved quite different IQ scores in Raven's test from those in the other two tests. This low correlation between the results of the respective tests reinforces the idea that special knowledge plays a large part in the results of WAIS and Stanford-Binet, but not of Raven's Matrices.

10

THE STANFORD-BINET TEST

Most of the problems in the Stanford-Binet test were collected at the end of the 1930's in connection with the revision of Stanford-Binet-16, in 1937. The test was revised again in 1960, but no new items were introduced; on the contrary, a lot of items were screened out because one of the two equivalent test series L and M was scrapped.

Stanford-Binet consists of 17 tests, one for each year from age 2. The last test for a specific age is the fourteen year-old test. Of the remaining four tests, one is for average adults and three for superior adults. The test for small children and parts of the other tests include performance tasks. But broadly speaking Stanford-Binet is a completely verbal test. To describe it I have selected Test III from the series of three tests for superior adults in equivalent form M, from Stanford-Binet 37:

1. Proverbs
Here is a proverb, and you are supposed to tell what it means. For example, this proverb, "Large oaks from little acorns grow" means that great things may have small beginnings. What do these mean?
    a) Let sleeping dogs lie.
    b) A bad workman quarrels with his tools.
    c) It's an ill wind that blows nobody good.
2. Memory for sentences
Listen, and be sure to say exactly what I say: "At the end of the week the newspaper published a complete account of the experiences of the great explorer."
3. Orientation, direction
I drove south three miles, turned to my *left* and drove east two miles, then turned to my *left* again and drove three miles, and then to my *left* again and

drove one mile. What direction was I going then? How far was I from my starting point when I stopped?

4. Repeating nine digits

I am going to say some numbers and when I have finished I want you to say them just as I do. Listen carefully, and get them just right. (The numbers must be pronounced clearly and distinctly by the tester, at the rate of one digit per second.)

a) 3 – 7 – 1 – 8 – 2 – 6 – 4 – 9 – 5
b) 7 – 3 – 9 – 4 – 8 – 1 – 5 – 2 – 6
c) 8 – 5 – 2 – 9 – 6 – 3 – 1 – 4 – 7

5. Opposite analogies

a) Ability is native; education is . . .?
b) Music is harmonious; noise is . . .?
c) A person who talks a great deal is loquacious; one who has little to say is . . .?

6. Repeating thought of passage: Tests

I am going to read a short paragraph. When I have finished you are to repeat as much of it as you can. You don't need to remember the exact words, but listen carefully so that you can tell me everything it says.

"Tests such as we are now making are of value both for the advancement of science and for the information of the person who is tested. It is important for science to learn how people differ and on what factors these differences depend. If we can separate the influence of heredity from the influence of environment we may be able to apply our knowledge so as to guide human development. We may thus in some cases correct defects and develop abilities which we might otherwise neglect."

11

WAIS

The WAIS is the most popular intelligence test in the USA after Stanford-Binet. The first form of this test was published in 1939. David Wechsler, who was Chief Psychologist at a large hospital, considered the problems in Stanford-Binet much too childish for adults. He writes: "Asking the ordinary housewife to furnish you with a rhyme to the words 'day,' 'cat' and 'mill,' or an ex-army sergeant to give you a sentence with the words 'boy,' 'river,' 'ball,' is not particularly apt to evoke either interest or respect."

Perhaps it is natural for an army sergeant to think that peaceful occupations such as "the boys are playing ball down by the river" are silly, and perhaps he would prefer to answer the question "In what way are a banana and an orange alike?" (a problem from WAIS).

The Wechsler Adult Intelligence Scale, WAIS, comprises a verbal section and a performance section. The verbal section is divided into six sub-tests with six different types of problem. In each sub-test the problems are arranged in order, from easier to more difficult. I will give examples by quoting the easiest and the most difficult problems in each sub-test:

1. *Information*
Question 1: What are the colours in the American flag?
Question 29: What is the Apocrypha?
2. *Comprehension*
Question 1: Why do we wash clothes?
Question 14: What is the meaning of the proverb: "One swallow does not make a summer"?

3. *Arithmetic* (oral test)
Question 1: If you have three books and give one away, how many will you have left? (Time limit 15 seconds.)
Question 14: Eight men can finish a job in six days. How many men would be able to finish the job in half a day? (Time limit 120 seconds.)
4. *Similarities*
Question 1: In what way are oranges and bananas alike?
Question 13: In what way are flies and trees alike?
5. *Digit span*
Question 1: Repeat in the correct order the digits I am going to read out: 5–8–3.
Question 28: Repeat backwards in the correct order the series of digits I am going to read out now: 5–8–3–3–4–6–8–9–4.
6. *Vocabulary*
Question 1: What does bed mean?
Question 40: What does travesty mean?

These six verbal sub-tests are followed by five performance sub-tests.

7. *Kohs Block Design Test*
This consists of blocks which have to be put together to form a surface with the same pattern as a picture the testee is given to look at. The sides of the blocks are painted in different colours. There is a total of ten patterns to work from. Each pattern has to be completed within a time limit, which is 60 seconds for some of the patterns and 120 seconds for the rest.
8. This sub-test comprises eight series of different pictures. A series of pictures has to be arranged in sequence to tell a story in a natural way

Problem 1: Three pictures of a bird's nest, which have to be arranged to describe the event: a bird feeding its young.

Problem 8: Six pictures representing a man taking a taxi.

9. Four simple jigsaw puzzles have to be made. Each puzzle consists of five to ten pieces. The finished puzzles represent a hand, an elephant, a human being and a face in profile. the hand and the elephant must be completed in three minutes, the human being and the profile in two.

10. Nine different symbols have to be paired with nine numbers, according to a key on which the numbers are shown paired with the symbols. Time limit: 1½ minutes.

11. The testee is shown 21 different pictures from which something is missing. He must say what is missing from each picture. Time limit: 20 seconds per picture.

Problem 1: The picture shows a door. The door handle is missing.

Problem 21: The picture shows a girl. The girl's eyebrows are missing.

In student groups the verbal section of WAIS has a correlation of 0.50 with their grades. The correlation between the performance section and school grades is considerably lower.

In large groups, which are heterogeneous as regards intelligence, the correlation between the WAIS and Stanford-Binet results is 0.80 to 0.90. In a carefully controlled experiment WAIS and Stanford-Binet tests were given to 52 young whites from an approved school. According to Stanford-Binet the average IQ of the group was 100.5; according to WAIS, 94.4. According to the Stanford-Binet results the group was very heterogeneous: many had very high IQs and many very low IQs. According to WAIS the opposite was true: the group was very homogeneous. The standard deviations were 17.3 for Stanford-Binet and 11.7 for WAIS. "We may speculate as to which of the descriptions of intelligence in the group studied was the most accurate," writes David Wechsler.

The Wechsler Intelligence Scale for Children, WISC, is a test for children, consisting of the same type of problems as WAIS.

12

INTELLIGENCE AND THE BRAIN

In the Anglo-Saxon literature people who in our terminology are psychologically disturbed are called mentally defective. Those with an IQ below 25 are also called idiots and those with an IQ between 50 and 70 imbeciles. An English expert on mental defectives, Professor A. F. Tredgold, has said that almost all idiots and imbeciles are suffering from brain damage which is visible to the naked eye. In general the brain is smaller than the average; its convolutions are less complex and often markedly irregular and abnormal and it often shows extreme deviations from the norm in development.

Autopsies on idiots and imbeciles generally show a brain which weighs 20% less than the average for that age. The grey matter is also found to consist to a great extent of scar tissue and accretions.

The American author Lashley destroyed parts of the brain to differing extents in 37 rats and then tested their ability to learn to find food in mazes of different degrees of difficulty. As expected, he found that the greater the damage the worse the learning capacity. Differences in performance were small between normal rats and brain damaged rats in the simplest mazes, but increased, the more difficult the mazes became. He also found that the position of the brain damage was insignificant; it was only the percentage of brain tissue destroyed which mattered.

Since the least intelligent human beings have damaged brains, one would expect the more intelligent of the unintelligent people also to have damaged brains, but not damaged to the same degree as those of idiots and imbeciles, which on average weigh some ounces less than the European average. J. A. Fraser-Roberts studied the siblings of 562 people whose IQ was between 30 and 68.

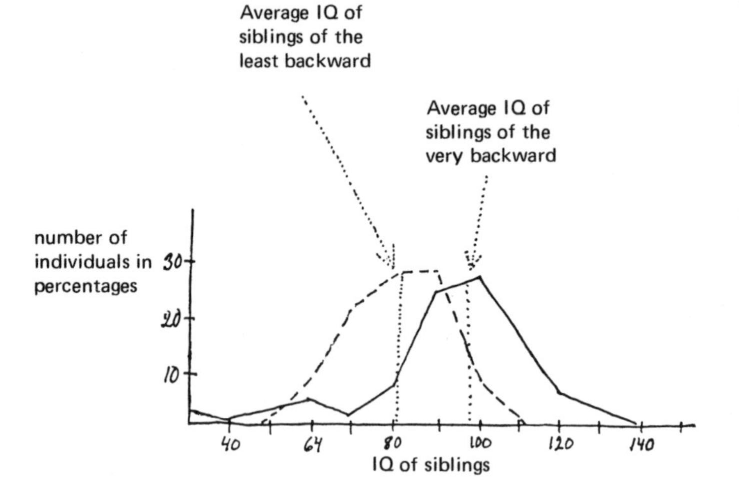

Figure 29. The result of Fraser Roberts' study of the IQ of 562 siblings of backward individuals (IQ between 30 and 68). The backward people were divided into two groups: those with the lowest IQ in one and those with the highest in the other.

He divided them into two groups. He found that on average the intelligence of the brothers and sisters of those with the lowest IQ, the idiots, was close to normal. But the brothers and sisters of the whole group of the more intelligent backward people had an average IQ of 80. The graph in Figure 29 illustrates the IQ distribution in the two sibling groups.

In Galton's time, and even up to the present day, a popular belief has existed that Europeans have better brains than people of other races. Most anthropologists working at the turn of the century were busy measuring head shapes and skull volumes to prove that Europeans had larger brains than other people. The usual textbooks state that Australian aborigines have brains which weigh a few ounces less, on average, than European brains. Vint, the anthropologist, took measurements proving that the brains of some African races were 10% smaller than those of Europeans. He also believed he had found that the quality of the nerve cells in the cerebral cortex of Africans was inferior by 15%. Another anthropologist, Gordon, found that the average annual growth of the African brain between the ages of 10 and 20 was only half as great as that of Europeans, though Heaven alone knows by what macabre means he may have checked his findings! The results of these studies should be taken with a large pinch of salt. In a similar study in the USA which gave similar results, a sharp assistant mixed up the white and African brains so that it was impossible to know which were white and which were Negro while they were being measured. When the investigation was carried out again under these conditions no average difference in weight could be demonstrated. The Rosenthal effect is probably particularly great in such a value-sensitive field of investigation as genetically based racial superiority or inferiority.

Anthropology was highly esteemed as a science until the Nazis were defeated in World War II. After that it was less highly prized. (Nowadays Swedish students are not allowed to study anthropology without special permission from the chancellor of their university.)

Now, it is true that women have smaller brains than men. Women's brains are smaller than one might expect from the fact that on average women weigh less than men. The results of intelligence tests show no corresponding variations, owing to the design of the tests. On the other hand, Western women are more dependent on authority than men, find it more difficult to give up a pattern once learned, and have a greater tendency to accept uncritically what is taught in school and at university. But these differences may be influenced by Western culture. The Rosenthal effect may also play a part. European culture is, after all, male dominated. Raymond Firth, a respected anthropologist, supplies an example. He writes that among the Bemba people in Northern Rhodesia recent contact with European culture has considerably increased the power of the father. Wife and children want European clothes and similar objects, and the father who can give his children such presents has more chance than before of retaining authority over his children. The fact that so many men now work in the copper mines has meant that the custom of working for the wife's family is beginning to die out, together with the habit of living in the wife's village. The power of the wife's relations has therefore diminished. European ignorance of this matrilinear system and European prejudices against it have also contributed to this. Audrey Richards says that young men believe that it is more English, and therefore more distinguished, to belong to their father's clan rather than their mother's, and this new custom has been encouraged by many missionaries.

There is a good deal of additional data to indicate that the size of the brain, as long as it is within the norm for the race in question, has no effect on the intelligence. Some Eskimo peoples have heavier brains than Europeans, and so have some Africans. The ethnic group, totally exterminated by white colonists, which once lived in Tasmania (they were hunted, their water was poisoned, etc., just as some "inconvenient" Indian tribes are being exterminated today in Brazil), belonged to a very special race, unlike any other. As far as we know, their brains were larger on average than those of Europeans.

The average European brain weighs 1,400 grams. A number of individuals, classified as geniuses, have had considerably larger brains. Turgenev, for instance, had a brain weighing 2,000 grams. Anatole France had a very small brain, weighing little more than 1,000 grams. A completely normal 46 year-old labourer had a brain weight of 680 grams, although his height was 170 cm and he weighed 70 kilos. Brain weight is correlated with body weight and height. Taller people generally have larger brains.

But if it is not the size of the brain which causes variations in intelligence, then it must be its quality. Do hyper-intelligent people have brains of better quality than others? I honestly do not think there is any data to indicate this. There does not seem to be a brain that is better than the "normal" brain. The normal brain seems to be so designed that the more data you give it, the more new data it can receive. There are no examples of learned professors who have learned so much that they could not learn any more.

Setting aside knowledge, interests, self-confidence and values, variations in IQ among the majority of a population arise from the fact that some are slower in their mental activities and others are quicker and that some have better memories than others. So one might think that some people have inherited brains in which the processes of nerve transmission are quicker. And it is possible that there are genetically based variations between different individuals in this respect. But for most people speed in mental arithmetic, for instance, has little connection with such neurological differences. The habit, the way or method of mentally manipulating numerical symbols and the existence of various psychological barriers in the form of inhibitions and associations, probably give rise to about 99% of the variations in the time taken for problem solving by different individuals. The neural processes work so fast that any inherent differences in the speed of these processes take place in hundredths and tenths of seconds; variations in different individuals' speed in solving test problems, to the extent that they influence the IQ rating, are in terms of seconds and tens of seconds.

As far as memory is concerned, we know that many idiots have incredibly vast memories and can repeat very long sequences of symbols.

To sum up, we can say that even if there are hereditary tendencies towards better brains or worse brains, these differences cannot be measured by intelligence tests because factors other than such qualitative variations play such a large role in the IQ score.

The relationship between the brain and intellectual capacity is obscure. We know that a series of electric shocks impair intelligence and we have proof that lobotomies reduce the IQ. But these are generalities. Neural surgeons have removed large parts of brains without affecting patients' intelligence. In one case, part of the right hemisphere, the left frontal lobe and parts of the left temporal lobe were removed without affecting the patient's capacity to read, remember and carry on a normal conversation. Sometimes operations of this type can result in a loss of sensation and the ability to move, but nevertheless do not always entail any impairment of the patient's intellectual functions. Data of this type have led the psychologist J. R. Kantor to conclude that it is unfortunate that intellectual psychological phenomena are still regarded as general functions conditional on some special type of nerve base structure, a view which undoubtedly illustrates the powerful influence of traditional ideas on our thinking. Kantor probably means that if a brain is normally developed there will be scarcely any difference in quality from brains of other normal individuals.

(There is a good deal of recently discovered data to indicate that the part the brain plays in thinking may not be what we have always imagined. For instance, when you think of an elephant, you often "see with the inner eye" a picture of an elephant. The picture may not be in the *brain*.

It is not impossible that the brain is "only" a system which receives sensory impressions and plugs them in to different types of function. The fact that human beings have the largest brain of all animals is adequately explained by the fact that the human being's hand muscles, neck muscles, tongue and lips require an infinitely larger wiring system than animals need for their functions. It is possible that human thought processes go on inside the brain, but it is not absolutely certain.)

13

NORMAL VERSUS ABOVE-AVERAGE

Can anyone with a normally developed nervous system solve the most difficult problems in intelligence tests, or is it necessary to have an above-average brain?

The answer to this question obviously lies in my setting you the most difficult problems in the three most important intelligence tests. If you can do them, either you have brains which are qualitatively better than normal brains, or else a normal brain is sufficient to enable you to solve this type of problem.

First comes the Stanford-Binet. I will give one of three tests for superior adults:

Problem 1: Vocabulary
What is meant by orange, envelope, straw, puddle, tap, gown, eyelash, roar, scorch, muzzle, haste, lecture, Mars, skill, juggler, brunette, peculiarity, priceless, regard, disproportionate, shrewd, tolerate, stave, lotus, bewail, repose, mosaic, flaunt, philanthropy, ochre, frustrate, incrustation, milksop, harpy, ambergris, piscatorial, depredation, perfunctory, limpet, achromatic, casuistry, homunculus, sudorific, retroactive, parterre.

Problem 2: Proverbs
Here is a proverb and you are supposed to tell me what it means. For example, this proverb, "Large oaks from little acorns grow," means that great things may have small beginnings. What do these mean?

a) A bird in the hand is worth two in the bush.
b) You can't make a silk purse out of a sow's ear.

Problem 3: Opposites

a) A rabbit is timid; a lion is . . .
b) The pine tree is evergreen; the poplar is . . .
c) A debt is a liability; an income is . . .

Problem 4: Repeating thought of passage: Value of life.

I am going to read a short paragraph. When I have finished you are to repeat as much as you can. You don't need to remember the exact words, but listen carefully so that you can tell me everything it says.

"Many opinions have been given on the value of life. Some call it good, others call it bad. It would be nearer correct to say that it is mediocre, for on the one hand our happiness is never as great as we should like, and on the other hand our misfortunes are never as great as our enemies would wish for us. It is this mediocrity of life which prevents it from being radically unjust."

Problem 5: Reasoning

I planted a tree that was 8 inches tall. At the end of the first year it was 12 inches tall; at the end of the second year it was 18 inches tall; and at the end of the third year it was 27 inches tall. How tall was it at the end of the fourth year?

Problem 6: Repeating nine digits

I am going to say some numbers and when I have finished I want you to say them just as I do. Listen carefully, and get them just right:

a) 5–9–6–1–3–8–2–7–4
b) 9–2–5–8–4–1–7–3–6
c) 4–7–2–9–1–6–8–5–3

So much for Stanford-Binet. In the most difficult problems in the WAIS 'Digit Span' test the figures have to be repeated backwards. Otherwise the verbal problems in WAIS are of the same type as in Stanford-Binet, with the exception of the sub-test on general information. The questions in this sub-test are knowledge questions on the lines of: "How far is it from New York to Paris?" In the performance section of WAIS all the sub-tests are so easy that anyone could solve them. It is the time limit which produces the variations in the frequency of solutions.

In the less well-established intelligence tests the problems generally regarded as most difficult are mathematical series. Example: 512, 32, 8, ?

If you are patient and very familiar with numerical symbols and the rules pertaining to them you can always solve problems like these, no matter how difficult they appear. In the last example, when half of a number in the series is multiplied by its whole, the result is the immediately preceding number in the series.

As I have mentioned, many psychologists regard verbal tests such as WAIS and Stanford-Binet as unsuitable for the measurement of intelligence because they call for certain knowledge and valuations. Raven's Progressive

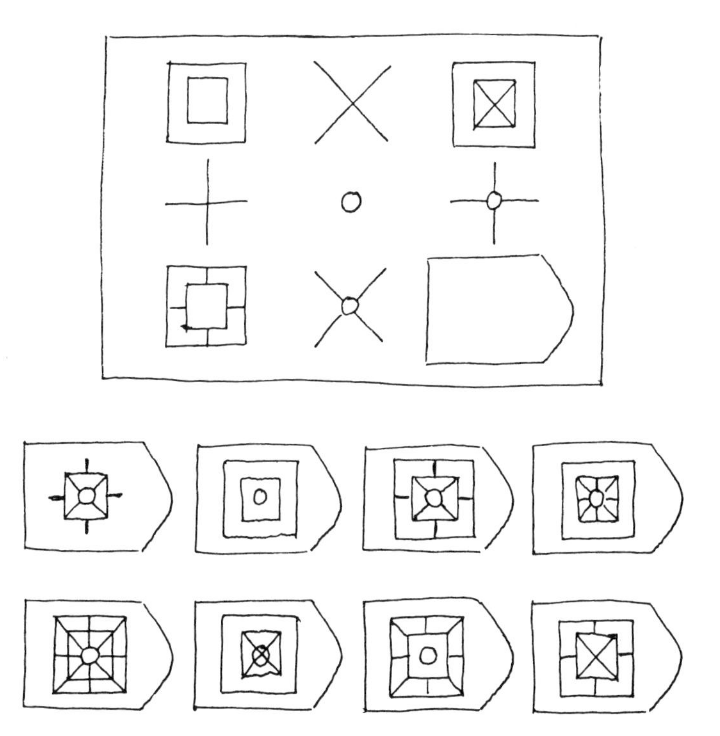

Figure 30. Problem from Advanced Progressive Matrices Set II: Intermediate.

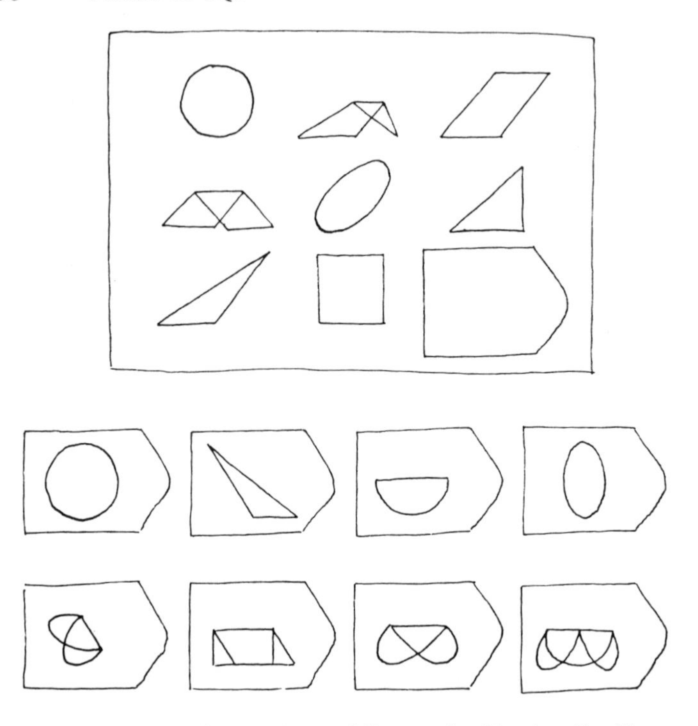

Figure 31. Problem from Advanced Progressive Matrices Set II: more difficult than the previous example.

Matrices superficially require no knowledge. By comparison with Stanford-Binet and WAIS, Raven's Matrices measure "pure" intelligence. Professor Raven has a number of Matrices intended for average people and other Matrices intended for the highly intelligent. In Figures 30, 31 and 32 I have reproduced three of the Matrices from Advanced Progressive Matrices Set II: one of average difficulty, one quite difficult and the most difficult of all. Raven's Matrices Test has no time limit and I think anyone who is sufficiently interested to study the Matrices until he has found Raven's theory which underlies them will agree that what these problems demand of a person is not a qualitatively superior brain, but patience, interest in the problem and a more or less systematic search for the idea behind the series of symbols in the "Flag."

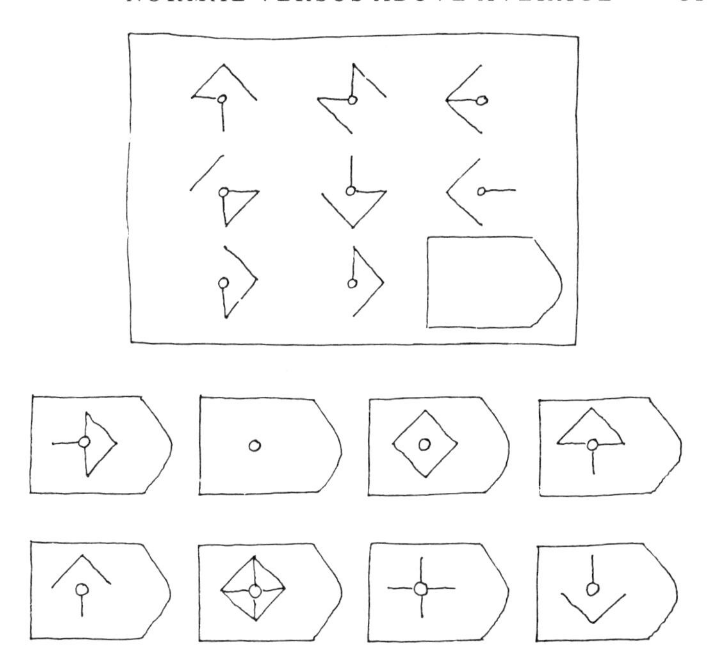

Figure 32. Problem from Advanced Progressive Matrices Set II: most difficult.

So as far as the commonest and best established intelligence tests are concerned, anyone with a normal nervous system and the necessary training and knowledge can solve even the most difficult problems. On the other hand, it should be clearly realized that it is possible to create a problem which takes takes an enormous length of time to solve, a truly subtle and complex logical pattern, and call it an intelligence test. The solution of these probably demands as much interest and as much thought as were expended on the problem.

14

MENTAL DEFECTIVES AND INTELLIGENCE

Mental defectives are closely associated with the concept of intelligence. One of the criteria of validity for intelligence tests is their ability to distinguish between mental defectives and normal people by their rating. (See the section below on various criteria of validity.)

The classification of mental defectives is often based on intelligence tests. According to the Norwegian educationist Magne Nyborg, the terminological conventions shown in Table I are valid in different countries:

Table 1. Terminological conventions in different counties

IQ	Term	Country
Under 25	Extremely backward	Sweden
Under 25	Idiot	USA, Norway, Denmark
25–50	Imbecile	USA, Norway, Denmark
25–70	Moderately backward	Sweden
50–70	Moronic	USA
50–70	Feeble-minded	Denmark, Norway
70–90	Remedial grade	Sweden

The classification, independent of IQ, of individuals as mental defectives is based on their ability to manage in life. Those who are unable to manoeuvre their way through everyday life without doing themselves injury belong to the lowest group, the idiots. For higher groups the classification is based on a general assessment of their linguistic ability, learning ability, ability to manage without help, and ability to play a useful part in production.

The classification scheme below is based on that proposed for the psychiatric section in the next (ninth) revision of the International Classification of Diseases.

Normal Variation
(Including superior intelligence.)
Mild Mental Retardation
Inclusion Terms
Feeble-minded
Moron
High grade defect
IQ 50–70
Moderate Mental Retardation
Inclusion terms
Imbecile IQ 35–49
Moderate mental subnormality
Severe Mental Retardation
Inclusion terms
Imbecile NOS
Severe mental subnormality
IQ 20–34
Profound Mental Retardation
Inclusion terms
Idiocy
Profound mental subnormality
IQ under 20

The boundary between conduct regarded as mentally defective and normal conduct is a shifting one. The classification is reasonably objective only in relation to a person who cannot manage without injuring himself. For instance, it is impossible to distinguish mental defectives from the mentally sick in most borderline cases. Moreover, many mental defectives are obviously also mentally sick. Many of the mentally sick also have a very low IQ and the average IQ score of the mentally sick is below that of the population as a whole.

For these reasons the percentage of mental defectives in the population is assessed differently in different countries. One British expert works on a basis of 0.4% of the population, another takes 0.2%. The great Scottish intelligence studies of 1943 and 1947 found oover 1% of the eleven-year-old schoolchildren studied to be mentally defective. In New York State there has been mental health legislation based on an IQ concept which, according to Professor Wechsler, might entail the authorities' institutionalizing 20% of the white population and a considerably higher proportion of the black population as mental defectives, if the law were applied.

The mental defectives in care at various institutions in England amount to about 0.1% of the population. In Sweden proportionally twice as many mental defectives are cared for in institutions. According to the Statistical Year Book for 1969, the number of the psychologically disturbed cared for in general institutions for this purpose was 15,608, which is about 0.2% of the population. Those classified as mentally defective have a very low average life expectation compared with the normal population. This is an additional cause of difficulty in calculating the proportion they constitute of the population.

The causes of mental deficiency are unknown in most cases. It is estimated that the cause is partially identified in about 20% of the cases which are so defective as to be classed as extremely backward (idiots or imbeciles). The largest known group consists of so-called Mongoloid idiots. They usually make up about 10% of all those classified. Next come individuals with the congenital metabolic malfunction, phenylketonuria, making up 1–3% of the total.

We know that most mental defectives were malformed at birth. Some of them were born prematurely. In general, the factors leading to foetal death and abortion may also lead to congenital defects in the child if it survives. Damaging factors which affect the foetal stage have a tendency to affect the central nervous system to a greater extent than other parts of the body. Mentally defective children, children deformed in other ways at birth, premature and stillborn children and abortions are more usual in the lower classes of society than the higher. Schizophrenia and other mental diseases also seem to be rather more usual in the lower income groups of a society.

In Galton's time (and to some extent even now) the predominant opinion among scientists was that congenital mental defects were genetically based. This was supported by data on the children of idiots and imbeciles, an abnormally high proportion of whom were themselves mentally defective. A common figure for the occurrence of mental defects in the relatives of idiots and imbeciles is 7–9%, as against about 1% for the rest of the population. It should not be forgotten that percentage figures such as these are dependent on how one defines "relative." If by relative one means brothers, sisters, parents and all the parents' brothers and sisters and their children, the percentage will be lower than if "relative" is taken to mean only parents and siblings.

It has also been observed that marriage between cousins is more common among parents of mental defectives than in the population as a whole.

These facts about mental defectives and relationships do not necessarily mean that genetic factors play a predominant role, or that they play any role at all. Mental defectives are not capable of feeding themselves adequately, of living healthily and avoiding physical injury (after all, this is one of the definitions of mental deficiency). So one might expect their children to be more often malformed and defective in various ways than other people's children. There are also cases where one of two monozygotic twins was a blind, microcephalic (abnormally small-skulled) idiot, while the other was completely normal.

Radioactive irradiation of the mother during pregnancy can cause mental deficiency in children. This was proved in population studies in Hiroshima and Nagasaki. An English study has shown that of 74 children whose mothers were X-rayed during pregnancy, 33 had poor health and 25 were born with malformations, particularly microcephaly. This type of radiation damage to the foetus does not appear to be hereditary. Studies of children of the radiation-damaged Hiroshima children (those who retained their reproductive capacity) tended to confirm this.

There is an above-average tendency for mothers of mental defectives to have been sickly before pregnancy.

A study by Klebanov demonstrates that psychological factors are also involved. He studied women who had been in concentration camps during World War II but had survived and then lived normally for some years. The frequency of congenital deformities in children born to such women long after their release from the camps was several times as high as that of the population in general. Specially common defects were Mongolism and microcephaly.

If the mother has mumps or German measles during pregnancy the child is quite likely to be defective. The Swedish authors Grönvall and Selander report on five deformed children among the offspring of 34 women who had had mumps during pregnancy.

Unsuitable food during pregnancy was found by Murphy in 1947 to be a further cause of congenital malformation. He calculated that 40% of mothers of deformed children studied by him had had insufficient or wrongly balanced food while they were bearing their children. Alcoholism in mothers is also suspected to be related to mental defects in their offspring.

One of the most important causes of mental deficiency is injury during delivery. This is common (affecting some 10% of new-born babies to a greater or lesser extent) and produces damage to the brain in the first place, since the head is the largest part of the body to pass through the pelvis. It is also the part of the body most immediately damaged by oxygen deficiency during labour. Norman L. Corah is one expert who has followed up a group of children exposed to oxygen deficiency at birth, to the age of seven years. There were wide variations in IQ, but these had a tendency to disappear as the children grew older. Presumably the effects of brain damage faded into the background, in relation to the differences in home environment and methods of upbringing.

Finally mention should be made of failed abortion attempts, hot baths, quinine, saffron, pills, physical intervention with bicycle spokes, teaspoons, etc. The occurrence of these is considerable among mothers of mentally defective children, according to Professors L. T. Hilliard and Brian H. Kirman.

As I mentioned before, the brains of the severely mentally defective individuals generally deviate greatly from the normal. The differences from the normal vary in a number of ways which make it improbable that special genes are the cause of the disturbances. On the contrary, it seems to be the rudiments of the normal nervous system which have been damaged. Another fact which points in the same direction was referred to in the section on intelligence and the brain (see Figure 29). The brothers and sisters of extreme imbeciles were more intelligent than the brothers and sisters of the more intelligent backward subjects. The brothers and sisters of the least intelligent backward subjects had an average IQ of 100 while the brothers and sisters of the most intelligent had an IQ score of only 80 on average.

Among mental defectives as a whole there are some who are easily type-classified. One such type is the Mongoloid idiot. Although we have found that the mothers of Mongoloid children are slightly older on average than the mothers of normal children, many authors consider that this variant of mental deficiency is genetically based. The number of chromosomes in Mongoloids has been reported as abnormal. The brains of Mongoloid children weigh less than those of normal children, but they are heavier than the brains of most other extreme defectives. Mongoloid children do not display extreme microcephaly, nor do their brains show the usual scar formations and deformations of the gyrus. They seem normal but not fully developed. So there is reason to suspect that the Mongoloid syndrome is the result of a defective hereditary tendency in the fertilized egg. But at the present stage of scientific discovery we cannot say if the defect in the hereditary factors arose during or after conception, or if it was there before and was therefore inherited.

Mental defectives have an above-average number of defects other than a low IQ. Many have motor handicaps, many are blind, a large number have inferior vision, most have inferior hearing and many are deaf. It is estimated that on average, deafness means a twenty point shortfall in IQ score by comparison with otherwise similar individuals, owing to reduced ability to learn the language. Twenty IQ points is a quite arbitrary figure in this connection. If no one sees to it that a person learns the most important ground rules for verbal communication, he will not learn enough to get any IQ points at all.

In most cases a suitable environment and suitable training can teach mental defectives to fend for themselves in life. For the few mental defectives who are most seriously damaged there is not much to be done, other than to try to make their life as rich in happiness as possible while it lasts. Often the most severely defective die after a few years. The most usual cause of death is disease of the lungs and airways. Their nervous systems cannot cope with the coordination of eating, swallowing and respiratory movements.

The retarded, on the other hand, are quite different. In a study made by S. A. Kirk in the 1950's in the USA, 81 mentally retarded children aged from 3 to 6 years were identified and studied over a period of from 3 to 5 years. The children were divided into four groups. The first group of 28 children were sent to a kindergarten in the little community where they lived with their families. The second group, of 26 children, lived in the same way as the first and did not attend kindergarten. The third group of 15 children lived in a home for the mentally retarded and went to kindergarten. The fourth group of 12 children lived in a similar institution but did not go to kindergarten. At the start of the experiment all the children had IQs between 45 and 80.

Of the 43 children who went to kindergarten, 70% showed an increase in their rate of intellectual development, i.e. their IQs increased. When the children were divided according to whether their backwardness was organic or not, those who had not been diagnosed as organically retarded showed the greatest increase in intellectual capacity. In the group of 15 children, 5 out of 8 non-organically defective children increased their capacity to the extent that they were discharged from the institution. But only one of the 7 organically defective children was discharged. On the other hand, half of the children diagnosed as organically defective also increased their IQ.

One side-effect of Kirk's study was that light was shed on the importance of the home environment. A closer examination of the home background in the case of the children in the study who lived at home showed that through their conduct (indifference or hostility) to the children some of the parents were aggravating the results of the mental defects.

The French phrase "idiots savants" refers to mental defectives who have certain intellectual capacities to an unusually high degree. The literature describes hundreds of mental defectives who have been above-average individuals in some spheres. One example is a brain-damaged 38 year-old man who after hearing two and a half normal pages of text read out once, could repeat them word for word. He was also an excellent pianist.

Strikingly often such descriptions in the literature concern individuals who had time data "on the brain" and could give the day of the week, the date and the year in all sorts of connections. A person with an IQ of 50 could give the day of the week for any date you chose between 1880 and 1950.

An individual with an IQ of 20 could immediately calculate the fourth and fifth root of three-figure numbers and give the product of very large numbers without a second thought.

An American study of the occurrence of "idiots savants" produced a total figure of 33 "idiots savants" at fifty-five institutions. Of these 8 were musicians, 8 mathematicians and 7 draughtsmen.

Finally I would like to mention a study published by Alvin R. Howard in 1966. He followed up mental defectives and other hospital patients who had been tested with Wechsler's memory test 15 years before. Briefly summarized, his findings were that continued institutionalization seemed to be an inhibiting factor in various types of intellectual functions. The non-organically defective person could in time become indistinguishable from his organically defective counterpart.

15

INTELLIGENCE AND GENIUS

It was Francis Galton who conceived the idea that the term "genius" could be applied to an IQ of 140–150 and above. On these terms, any lecturer or professor or accountant could be declared a genius.

In general, the word genius is used to describe people who have been highly aware, non-authoritarian and independent of contemporary attitudes, often creating by virtue of these qualities work which has expanded the horizons of humanity, both figuratively and literally. Leonardo da Vinci, Charles Darwin, Sigmund Freud and Albert Einstein possessed such qualities.

On the other hand, there are many people who have always been regarded as geniuses, but who seem to have been less perceptive. Plato is a good example. He quite seriously regarded what we see, hear, smell, taste and feel as less real than the ideas and concepts with which we are inculcated. He never appreciated that our ideas, our thought categories, are in the first place the products of our speech. In his approach to true reality Plato showed a disquieting affinity with the religious fanatics who insist that the true reality is somewhere else, in another time.

Since intelligence tests principally measure a man's ability to pass intelligence tests, it is hardly appropriate to describe the people with the highest scores as geniuses. The results of intelligence tests say nothing about the ability to transcend the barriers of contemporary thought and the ability to find new and brilliant solutions to old problems. It is in the very nature of the intelligence test that the testee must offer solutions to problems already solved and also that the problem must be solved according to certain given norms. Of course, one could say that a person with an IQ below 120–130 will never be called a genius by posterity, but beyond the normal senior school and student level the variations in scores cease to bear any reference to the idea of genius. What happens is that a person with a very high IQ has displayed his social indoctrination into a special way of thinking and using language which is not always compatible with the ability to go outside the "correct" norms and create completely new work and theories. During their lifetime many geniuses were regarded as cranks by contemporary authorities. Obvious examples are Copernicus, who claimed that the earth was a planet travelling round the sun, Bruno, who was burned at the stake in 1600 for delivering university lectures on the theory that the universe is infinite and the earth is not the only world in the universe, and Einstein, who at the start of his career was regarded as a fool.

In the 1920's people with an IQ over 150 were called geniuses. Professor Lewis Terman, the man behind the Stanford-Binet test, initiated a gigantic study of individuals with an IQ of genius level. He collected information from all Californian schools on their most gifted pupils, whom he then tested by means of intelligence tests and other tests. On the basis of the test results a total of 1,528 children with an IQ above 140 were selected (the group included 62 children with an IQ between 135 and 140). The average score was 151 and at the top was a girl with an IQ of 200. These children represented the most intelligent level of the population of California.

A detailed case history was written about every one of the children, from their birth, nursing, earliest development, illnesses, nervous symptoms, to the time when they learned to read, etc., etc. 783 of the children underwent a thorough medical examination and almost the same number were measured by anthropological methods in order to obtain information on the skull shape, brain size and racial characteristics.

The results? For a start, it was observed that 31% of the children in this hyper-intelligent group had parents from the highest social group: directors, academics, land-owners. This group made up only 5–10% of the whole population. Only 7% of the hyper-intelligent children had fathers who were completely unskilled labourers, whereas this labouring category made up about 40% of the whole Californian population.

There were more books in the parental homes of the hyper-intelligent children than in the average home: an average of 328. Another difference was that the parents of the hyper-intelligent had 4–5 years more schooling behind them than the average parent.

An unusually high percentage of the hyper-intelligent were first-born or only children. They had been breast-fed rather than brought up on the bottle to a greater extent than ordinary children. Their mothers had been healthier during pregnancy than other mothers.

On average, the hyper-intelligent children had learned to walk a month earlier than other children. They had learned to speak, on average, 3.5 months earlier. They had rather better health, were muscularly a little stronger and a little taller than other children. They generally slept an hour or two longer per day than the average child.

The hyper-intelligent children read roughly twice as much as their contemporaries. Many had learned to read before they began school. Their schooling in the lower classes seemed to be without significance to their development in intellectual respects. Terman worked out the correlation coefficient of their level of knowledge and the number of months spent in school and obtained a small, but negative coefficient of – 0.10. The longer they had spent in school, the less they knew. However, a correlation coefficient of 0.10 is so small that it would be wrong to draw any conclusions from it.

The progress of the hyper-intelligent was followed up by sending them a form every fifth year with questions to fill in and to be returned to Terman. They were also given an intelligence test approximately every tenth year. In the first follow-up intelligence test, carried out six years after the initial tests, the boys' IQ had dropped 3 points on average, whereas the girls' had dropped by 14 points. In 1945, when the hyper-intelligent were about 35 years old, Terman published data about their development in "The Gifted Child Grows Up." The data were summarized again when they were 45 years old. At that stage the hyper-intelligent were on average less intelligent than in 1922 by a full 17 IQ points. It might be interesting to try to analyse the causes of this sharp decrease in intelligence.

Firstly, most of the hyper-intelligent had an IQ of just over 140 when they were selected. Far fewer had more than 150 points. Still fewer had more than 160 points, etc. Since the reliability of the intelligence test results is about 0.7 for a group of this degree of homogeneity, it was often chance which decided who should be included in Terman's group. One girl who had an IQ of 142 on the first test was included. It is very probable that in another test she would have got 134, and 137 in a third test. This means that the "true" or "real" IQ level for most of those tested was slightly lower than Terman's figures. All those who "in fact" had 142 but received only 138 points in the test were included.

Secondly, most of the hyper-intelligent had long since finished with school and university education by the age of 45. We know that the average IQ for large groups always drops when schooling ends, or when they begin to adapt to their professional work, which is unlike the type of problems which arise in intelligence tests.

Thirdly, the hyper-intelligent were not tested by the same methods in all the follow-up tests. The highest correlations between different tests lie between 0.80 and 0.90. Fresh intelligence tests and revisions of old tests are "more difficult" than the older tests, because the intelligence level of the whole population increases with time and because the average IQ of the majority must always be precisely 100.

This goes hand in hand with the fact that the highly intelligent children of 1922 came to a great extent from the upper class. The advances in school education, and radio and World War II largely levelled out the differences in opportunity for intellectual development between different social groups. The average level of intelligence in the USA increased by about 10–20 IQ points between 1922 and 1955. We know this partly from comparisons made between the results of the Army Alpha in World War I and the results of the Army General Classification Test (AGCT) in World War II.

If we take these factors into account it is probably

closer to the truth to say that the hyper-intelligent group did in fact increase in intelligence between 1922 and 1955. This is what one might expect from the fact that their intellectual self-confidence and interest in intellectual occupations must have been greatly increased by their awareness of belonging to an elite group. Their great experience of intelligence test was also likely to give them a higher IQ.

As adults the hyper-intelligent men were generally academics, directors and craftsmen. 60% of the hyper-intelligent women were housewives and those who worked professionally were generally office workers. It is a fact that the hyper-intelligent women chose office work before teaching to a far greater extent than women college students in general.

About five times as many of the hyper-intelligent took a doctor's degree than college students in general. There were more doctors of laws than doctors of philosophy and medicine.

In relation to the theory of heredity it is interesting that the children of these hyper-intelligent people had an average IQ of 133. This is almost exactly the same rating as the average for their parents at the age of 45 (134). This is not in accordance with the genetic doctrine, according to which regression should have reduced the children's IQ to 117, the median value between the norm of 100 and the parental 134. The percentage of mental defectives among the children of the hyper-intelligent was about half the percentage for the population in general, while the incidence of mental diseases among the relatives of the hyper-intelligent was only half the incidence found in the population as a whole.

If the usual intelligence tests measure the ability to solve the problems of which the test consists, one might expect the difference between those who had an IQ of 150 and those with an IQ of 170 to be minimal with regard to their performance in life, and this was in fact what Terman and his colleagues found when they compared those with an IQ over 170 with the rest of the group.

Another rather interesting comparison was made, between the 150 "most successful" and the 150 "least successful" of the hyper-intelligent. The IQ variation between the groups was small. The average IQ in the successful group was 155, as against 150 in the less successful group. "Successful" is defined as "having a profession with a high status," "having produced books or works of art which were generally admired," etc. The intelligence of the least successful group had dropped more steeply since childhood. Separations were more usual in the parental homes of the less successful group and among themselves separations were twice as common as in the most successful group. Moreover, the less successful group's parents had a lower average education and belonged to lower social groups. In the least successful group the individual's self-confidence was markedly lower, according to measurements with attitude scales.

16

INTELLIGENCE AND ENVIRONMENTAL FACTORS

People from lower social groups have a lower average IQ than people from higher social groups. The differences are great. In the USA they are on average greater than the average differences between Negroes and whites. Approximately 30 IQ points separate unskilled labourers from academics, directors and other highly qualified categories. Both American and Swedish studies show that office workers have higher IQs than artisans, who in turn have a higher IQ than workers in agriculture, forestry and fishery, to take just a few examples.

Of the 12 million men tested with the AGCT, 18,782 whites were selected as material for a professional group study. The 18,000 were divided into different groups according to their jobs, the average IQ of the groups was calculated and also the IQ variation in each group. According to the results, accountants, lawyers and engineers had the highest IQ, with an average of 128. Agricultural workers and labourers were at the bottom, with an average IQ of 95. The IQ of the highest category varied between 94 and 157. The heterogeneity was greater in the lowest categories, where the IQ varied between 24 and 149.

Apart from the psychological climate in the parental home, of which more later on, the IQ is determined by the length and quality of schooling. The correlation between these factors and the IQ is difficult to calculate, owing to the difficulty of assessing the quality of teaching. In a qualitative assessment of school education for 19 pairs of monozygotic twins the differences in this respect were judged by five independent assessors. The correlation between IQ variations and variations in school education (educational advantages) was 0.79, one of the highest to appear in intelligence research. Professor Torsten Husén's studies and follow-ups of a large group of state school children in Malmö showed an increase of 11 points in IQ for those leaving high school, as compared with the rest who had been at the same IQ level in the fourth class of their state school. Those who had begun work at the end of their state school life showed a drop of one point in IQ.

We know that a child with a certain IQ at seven will spend less time at school, and at worse schools, if his parents belong to a lower social group than if he comes from a higher socio-economic background. This means that IQ variations between social groups have a tendency to increase as time goes by. However, this tendency is balanced by many factors, of which the most important is the growing democratization of the school system and the increase in the length of obligatory school attendance.

It has been clearly proved that there is a strong connection between the parents' behaviour towards their children and the child's IQ. An important paper by Dr John Blomqvist has shown that Swedish schoolchildren who had a more than usually harmonious home environment and school situation also had the highest IQs. Goulet and Mazzei at West Virginia University studied the effect of an anxious basic attitude to life on the capacity to learn. They came to the conclusion that those who were most anxious in their general attitude found it most difficult to take in words and concepts. Because the constant learning which takes place at home and at school plays a decisive role in the level of the IQ, many people are handicapped. Those who grow up in an environment where they have reason for anxiety, where emotional security is lacking and rigorous demands are made, coupled with threats of punishment for failure, generally find it more difficult to learn. This in turn leads to less appreciation on the part of the parents, diminished self-confidence, poor school results and lower IQ.

An American study of adopted primary school children in the 1950s showed a relatively high correlation between the children's IQ and another aspect of the parental attitude: ambition for the child. The correlation was 0.30. The variations in IQ between social groups seem to depend mainly on differences in education, differences in home background, differences in pre-natal environment (environment in the womb) and differences in the attitudes of parents and neighbours. Environmental differences are often difficult to define. An example of a rather unusual environmental factor appears in the following extract from an article in *Sydsvenska Dagbladet*, March 1970:

> "Children in slum areas throughout America are as keen on sweets as other children in the world. But their parents are so poor that they are unable to buy sweets for the children. To satisfy their hunger for something which does not simply taste of food, about 10,000 American children between the ages of 1 and 6 eat old, flaking, lead paint every year. According to an article in the *Wall Street Journal*, countless children contract lead poisoning every year in the United States, leading to blindness, speech defects, deafness and backwardness.
>
> "It is impossible to collect correct figures on American children with lead poisoning, but the experts estimate that 250,000–400,000 children in America suffer from it.
>
> "These cases occur almost exclusively in the slums. The manufacture of lead paint stopped about 30 years ago. Few buildings put up after World War II have paint containing lead. The so-called residential districts of Harlem and other ghettos are generally over 30 years old and the paint there can be fatal to the hungry children.
>
> "The authorities are doing little to prevent lead poisoning of the very young. The symptoms of poisoning are roughly the same as those of chills and sickness. Urine tests seem to be the best method of detection. But laboratory tests take so long that the child is damaged for life by the time the examination has shown that lead poisoning has occurred.
>
> "The health authorities know the risk and do nothing. What can they do? There is a real risk that the householders would rather abandon their old hovels than be forced to paint or maintain them. It's

better to have houses in a poor state than houses which fall down altogether.

"But New York is beginning to wake up. A law has been introduced providing that the householder must repaint his house within five days of detection of a case of lead poisoning. If he refuses to repaint, the authorities send painters to the house and the bill to the owner.

"Reaction against the authorities' arrogance in the face of the risk of lead poisoning for children is growing. The poor families can't do much -- they are so poorly educated that they scarcely recognize the danger.

"In Harlem a group of young people have recently been making urine tests on 104 children, aged from 1 to 6. Thirty of the children, that is one-third of them, were so severely affected by lead poisoning that they were damaged for life."

There is cause for scepticism in relation to the statistics in the article. On the one hand, this is an article in a daily paper; on the other hand, these are data reported from some earlier period (there is a big time gap between the report and the source); also the article was written at a time when criticism of American community life had a positive value for the mass media. At the same time there is no doubt that the natural destruction arising from industrialization and the pollution of the environment, such as lead in paint and in vehicle exhausts, is more damaging to the lower than to the upper classes. The children of academics and businessmen are more likely to grow up in garden suburbs with fresh air, far from street crossings, traffic jams and factory chimneys.

Another of the factors which explains the average IQ variations between the different social groups is connected with the pre-natal environment. Here is a summary of a study carried out by Hilda Knobloch and Benjamin Pasamanick: in a study of mentally defective children born between the years 1923 and 1948, and admitted to Columbus State School, they found that significantly more had been born in the winter months of January, February and March. Since the third month following conception is known to be the period during which the cerebellum of the unborn child is formed, any damage occurring at that time may affect the intellectual functions. The months when this might happen are June, July and August, the hot summer months. If pregnant mothers reduce their food intake, especially proteins, at that time, their unborn children will be damaged.

If these factors are significant to IQ, one might expect individuals from slum areas to reduce their food intake at the hottest time of year, since there is less air-conditioning there than in other areas. Also, slum mothers have to stay in the hot city, while the higher classes of the community move out to the country.

It should be added that Swedish studies have also confirmed the greater incidence of mental defectives among children born in the winter months. Dr. Gösta Berglund in Uppsala found this to be true in a group of 237 individuals of less than average intelligence. For individuals from higher social groups this factor seems insignificant. Repeated studies show that the frequency of IQs of about 100 and above is unconnected with the season of birth.

We know that women from the lower social groups live on less adequate food than other women. This is also likely to increase the percentage of children with more or less defective brains in these social classes. In 1955 Ruth F. Harrell published a study of a group of pregnant women from a low social group. Some of the women were given vitamin supplements in the form of pills, while others were given neutral pills. Tests on their children at the ages of 3 and 4 showed that those whose mothers had received supplementary vitamins had a significantly higher IQ.

The circumstances of birth are among the environmental factors connected with IQ variations between different social groups. Birth involves great stresses on the infant. A study of a group of new-born babies showed blood in the spinal fluid of between 7 and 13% of the total. This means that the infant's central nervous system was damaged during delivery. To the extent that some general differences exist in the methods of delivery used in the lowest social groups as compared with higher social groups, these are reflected in the average IQ variations.

However, there is no doubt that differences in the home environment and schooling are responsible for most of the IQ variations between the social classes. Differences in early acquired attitudes are probably significant to scholastic success and therefore indirectly to the IQ. Some people are more rigid, dogmatic and conformist than others. They have a tendency to accept assertions (especially if they come from "up there," from the papers or television) and circumstances without opposition and also, unfortunately, without criticism. They also have a tendency to cling to an acquired behaviour pattern or pattern of problem-solving, even if it does not work well in a new situation. Group-pressure is a useful term in this connection. Pressure from the group to behave in conformity with its norms is accepted to a greater degree by the lower groups. The conformists are more emotionally inhibited, have a tendency to show less direct aggression (storing up their anger instead of letting their aggression overflow before it gets too violent) and are more anxious. The conformists are more authoritarian, more conventional and more bound by the prevailing moral standards than others. Studies have shown that the higher social groups have greater self-confidence and greater tolerance in moral, religious and sexual matters. Members of the lower social groups are more likely to be conformist.

There are significant IQ differences between conformist and non-conformist individuals. The attitudes which make an individual conformist are, from all the evidence, laid down very early, perhaps even in the first two or three years of life.

An important factor here is intellectual self-confidence: confidence in one's own ability to think and in one's own judgment in relation to that of others. A child brought up from the start with the admonition "Don't think for yourself, others think better than you," or "Don't interfere, you don't understand," is hardly likely to become an intellectual giant.

In his book "Intelligens och Tänkande" (Intelligence and Thinking) the Danish psychologist Dr Kaj Spelling reported on an interesting experiment in conformity:

> "Richard Crutchfield carried out some psychological experiments relating to conformity at the Institute of Personality Assessment and Research in Berkeley in 1955–59. He defined conformity as follows: if an individual has to express his views in a group situation and his own private opinions are in opposition to those of the group, he is in a conflict situation. There are two alternatives for the individual: 1) he can stick to his own opinion and thus make himself independent of the group's views, or 2) he can change in favour of the group's opinion, and this is conformity. Crutchfield's studies, briefly, reached the following conclusion:
>
> "The subjects were tested in groups of five without knowing anything about the purpose of the experiment. They had to express personal assessments in a series of different situations, e.g. decide which of two geometrical figures was the larger. The five people sat in booths separated from each other and had an instrument panel with several knobs in front of them by means of which they gave their answers. The subjects were not allowed to speak to each other but they could *apparently* communicate by means of the knobs. Each one signalled his answer in turn, and could read off the other's answers on his instrument panel. The sequence of replies was changed so that everyone was given a chance to answer first while the others awaited their turn.
>
> "That was how the subjects saw the situation, but they were being deceived by the conductor of the experiment. The answers of the other subjects which could be read off on the instrument panel were not their actual answers, but were signalled by the conductor of the experiment from his switchboard. In reality there was no connection between the five subjects. In this way the conductor was able to confront all the subjects with the same number of false 'group' norms.
>
> "The experiment was carried out on 600 indivi-

duals of different ages with different education, profession, social group, intelligence and personality. The results show that conformity varies considerably, i.e. there are great differences in the subjects' individual ability to resist what they experience as group pressure.

"A group of 50 officers was the most conformist. They achieved a value of 33% on average (the scale ran from 0 to 100%). After that came 53 second-year university students, with an average of 26%, 30 senior engineering students with 20%, and the most independent were 45 industrial researchers with a value of 14% on average.

"For the female subjects the average value was highest for 80 second-year university students, with 38%. 24 older students got 32% and the lowest value came from a group of middle-aged participants from a so-called Prestige Women's College, with an average of 22%.

"These results are fully in agreement with an assessment of the creative potential of the various groups, says Crutchfield.

"There is a significant difference between the degree of conformity in men and women, the latter being more constrained. A direct comparison can be made between the two groups of second-year students, in which the male and female students obtained average values of 26% and 38% respectively."

The myth of the unintelligent rustic

The research data are unanimous as regards the difference in IQ between children living in the country and children from the cities. Country children have an average IQ 10 points lower than that of city children. The difference is almost imperceptible in the lowest age groups but increases from year to year as the child grows older.

For a long time attempts were made to explain this in accordance with Galton's theory that the brightest people were attracted towards the cities and the congenitally dimmest stayed in the country out of sheer stupidity. But the results of studies have made psychologists more cautious in their reliance on this so-called migration theory. The children of Negro families who had moved into town and of New York Negroes were tested and divided up according to the number of years they had lived in an urban environment. Those who had lived there less than a year had an average IQ of 81, those who had lived in New York between 1 and 2 years had 84, those who had lived there between 3 and 4 years had 87 and those who had lived there more than 4 years had 91. Those who were born in New York had an average IQ of 93.

But the migration theory has not faded away altogether. Some studies show that the percentage of those with high IQs who have moved to the cities is higher than the percentage of those with low IQs who have moved. But it seems quite natural that this should be so. The ones with the highest IQ are those who have educated themselves to be, for instance, engineers, accountants and clerks. Those with the lowest IQ are agricultural workers. The need for accountants, clerks and engineers in the country is very small indeed. Whether it was genetic or social factors which made country children educate themselves to these professions, cannot be decided on the basis of this data. It remains a matter of faith.

If the IQ is principally dependent on the environment one might expect children growing up with worse than average educational opportunities not to have IQs which remained constant year after year, but to have steadily decreasing IQs, just like children from the country. And this has proved to be so. In the 1920s H. Gordon, an English schools inspector, discovered that the IQs of a group of children who lived with their parents, who were either bargees or wandering gypsies, diminished with the passing of the years. The bargees' children went to school only once or twice a week and otherwise lived with their often illiterate parents on the barges. The younger they were, the closer their IQ came to the normal 100. The correlation between their IQs and their ages was calculated and the figure of –0.76 was arrived at. The gypsy children had somewhat better opportunities for schooling and the corresponding correlation for them was –0.43.

Another telling example comes from tests on 3,252 children in 40 schools in isolated mountain districts of Tennessee in 1940. Ten years earlier a corresponding group of schoolchildren from the same villages had been tested. In the intermediate period up to 1940 significant economic, social and educational changes for the better had taken place. The average child in 1940 had an IQ score about ten points higher than the average child ten years earlier. The same test was used on both occasions.

Their IQs ten years earlier had not remained constant but had decreased at the rate of about two points per year (this figure was obtained by measuring schoolchildren in all six state school classes on each occasion). This reduction of IQ with age persisted in 1940, despite the rise in the average level of intelligence. The reduction means that the child's home environment, or school environment, or both, provided less adequate stimulus for its intellect than the intellectual stimulus offered by home and school in the USA as a whole.

17

INTELLIGENCE AND POSITION IN THE FAMILY

The size of families and IQ are related in two ways. On the one hand families from the lower social classes generally have more children than other families and on the other hand children with fewer brothers and sisters are more intelligent than those with a lot of brothers and sisters, because brothers and sisters communicate in ways which differ appreciably from the language and spirit of intelligence tests.

Table 2 summarizes the results of Scottish Surveys' partial study of 1,100 children from families of varying sizes. The fact that children from large families are on average less intelligent than others is also connected with the expense of keeping children at school. It is probable that in families on the same socio-economic level, the children from families with the smallest number of children stay at school longest. Children from large families also more often supplement the family income by working at an early age.

"Only" children and first-born children have their need for communication satisfied by adults to a greater extent than the others. Owing to the greater dissimilarity in the environment of adults in relation to children, contacts with adults occur via a communication system in which conventional verbal symbols play the major role. Communication between children is not quite the same. Children in the same family of brothers and sisters develop their own language, have their own jokes and allusions, which are often incomprehensible to their parents, and in any case have little to do with the conventional written and spoken language.

Of the 23 American astronauts who first went into space, 21 were first-born children and many were only children. Within the family the eldest child has been found to be more intelligent on average than the second child, who in turn is more intelligent than the third child, etc. The probable reason for this is the communication factor I have mentioned, plus the fact that the parents generally have greater ambitions for their first-born and expect more of him. There is good evidence to suggest that the first-born adopts his parents' moral standards and maintains tradition to a greater extent than the later children. The rebels in the family are generally among its younger members. The younger children are generally left more in peace to develop in the direction they choose and are often given greater freedom to go their own way in life. Their parents make less demands on them and give them less responsibility.

Table 2. Relationship between the number of children in the family and their average IQ. Figures from Scottish Survey 1947: a random sample of 1,100 11 year-old schoolchildren.

No. of children in family	IQ of children in family
1	113
2	109
3	105
4	101
5	96
6	91
7	93
8	91
9	91
10	95

Most of the environmental factors discussed here are relevant only to IQ variations between large groups of people. For the individual the most important environmental factors are different. The most important of all is the attention of parents or nearest relations. Admiration is the form of attention which is most sought after, by children as well as adults. Children are small people, who operate on the same fundamental psychological basis as adults. In the absence of admiration the child attracts attention to himself in the form of disapproval, physical punishment and sympathy. Behaviour which attracts no attention at all is gradually abandoned. Behaviour which attracts attention in one way or another is consolidated. Old ideas on upbringing would have us believe that punishment guided the child's behaviour in other directions. This is not quite true. If it had been true, most criminals would not be criminals. In general they have received much more, and stiffer punishment than other individuals. Every punishment is a form of attention and tends in one way or another to fix the child's conduct or attention on the sphere in which it was punished.

Many psychological and sociological experiments and studies have shown that there is a connection between the frequency of punishments in the home and the frequency of criminality. The reason for stealing, for instance, is generally not that the thief is not afraid of punishment. It is more probable that people steal because the socio-economic differences in a society are too great. There are also data which indicate that the emotional climate while growing up plays a major role in the frequency of criminal behaviour. A study covering 1,465 juvenile offenders showed that only 8% of them had parents who played with them, took them out on small excursions and so on. For the group of recidivist criminals the percentage was still lower, and for those who were guilty of the most serious crimes this percentage was lowest of all.

The American psychologists Bacon, Child and Barry studied 48 different ethnic groups, mostly from cultures without a written language. They studied two types of crime: theft and personal injury. Crimes of personal injury proved to be strongly associated with the relationship between children and parents. The authors found that the circumstances in childhood which led to a high frequency of personal injury among adults were as follows: a mother-child household with inadequate opportunities for the child to identify with the father in the first part of its life; a sleeping arrangement for mother and child giving rise to a strong relationship of dependence between child and mother; a subsequent socialization (training to function as an independent member of society) which tends to be abrupt, punishment-based and to create emotional disturbance in the child.

As regards theft, they found that a high differentiation of status in a society was one circumstance associated with a high frequency of theft. In the most thief-ridden societies and in the homes in which thieves had grown up there was a significantly higher frequency of punishment.

These facts about attention, punishment and relations with parents are highly significant to the discussion of environmental factors affecting IQ. They mean that if a child receives attention and perhaps also admiration when it shows that it has learned a new symbol, a new word, for instance, then the child will have a tendency to direct its behaviour towards learning new symbols. A child who receives no attention when he uses newly-learned symbols loses interest in this side of life.

If when little Peter crawls across the kitchen floor he finds a book and starts to turn the pages over and receives attention of an admiring nature from his mother, he will be inclined to turn the pages over the next time he bumps into the book. If little Peter has more and more of such experiences in the home, he will find books and letters and figures take on a pleasant coloration in school. And since the teacher will find little Peter interested in school work, she will begin to take a positive attitude towards little Peter which will reinforce little Peter's interest in intellectual activities, and so on, and so on, until one day little Peter is standing in the university lecture hall giving an inaugural lecture on child psychology.

Little Andrew, on the other hand, has received the kind of attention and admiration from his parents and friends which has made him prefer to play football rather than read books.

All in all, it is true to say that a child's general interests can be manoeuvred only to the extent that one can influence one's own genuine admiration for certain behavioural manifestations. By this I mean that it is scarcely ever possible to deceive a child and pretend that one likes and admires some of the things he does if one is not personally really interested in this aspect of life.

18

THE DIFFERENCE BETWEEN AN IQ OF 90 AND AN IQ OF 140

There are almost always wide differences in the previous environment and social background of an individual with an IQ of 90 and one with an IQ of 140. These differences will not be discussed here. Only the differences in intellectual performance will be dealt with. What do we then observe?

Firstly, the person with an IQ of 140 has a far greater aptitude for language and other symbol systems.

Secondly, the person with an IQ of 140 generally has slightly different values from one with an IQ of 90.

These differences will not be discussed now, either. Let us assume that the man with 90 and the man with 140 points have the same knowledge and values. This is unlikely, of course, but logically quite conceivable.

On this basis, what might be the differences between these two individuals? I will list the personality fields which might give rise to differences, but the order in which the various factors are listed bears no relationship to their relative importance.

Power of concentration

Obviously one must be able to devote one's attention to a problem in order to solve it. If the thinking process with which the problem is to be solved is interrupted the solution will be delayed. There are very wide differences between individuals in this respect. Some are interested in almost everything that happens around them. In a test situation the problem-solving activity of an individual of this type would be constantly interrupted, for instance by his sudden interest in what the conductor of the test is wearing, or the movement of leaves on a branch outside the window.

Distraction and its opposite, concentration, are connected with memory span.

Memory span

Some people can remember rows of intrinsically meaningless words or figures, while others find it difficult to keep them in mind. Some people can remember instructions given by the examiner or in the text of the various tasks; others quickly forget them and have to go back and ask or re-read the instructions. Where spatial problems are concerned the testee has to remember a two or three-dimensional figure, or parts of it, and then turn the figure within a given space and give the resulting position in the solution (see Figure 33). Another example of this type of problem is: "You have to collect 3 pints of water with a 4 pint can and a 9 pint can" (Stanford-Binet). Some problems, the problems which require the testee to repeat a row of figures read out to him, involve memory span alone.

Memory span is associated with retentive capacity.

Retentive capacity

Retentive capacity is the ability to retain information over quite long periods of time. Most intelligence-test problems depend more or less on this capacity for their solution.

Figure 33. Example of two-dimensional figure which has to be mentally reversed. The problem is to put a cross under the figure holding its arms out in the same way as the first figure in the series.

Calculation problems, for instance, are more difficult to solve for someone who has forgotten his multiplication tables and has to *work out* even the simplest multiplications. Other intelligence-test problems consist quite simply of questions concerned with information which the testee received at a much earlier time. An example of this is the WAIS question: "What is the population of the USA?"

Retentive capacity is associated with

Attitude towards the various ingredients of life

Some individuals find it extremely hard to learn and remember certain types of material, though in other connections they may have a very well-developed retentive capacity. Someone I know had great difficulty in learning the German prepositions which take the dative case, but he could repeat a song text of far greater length perfectly after hearing it only twice.

Rigidity

A person's rigidity affects his IQ because it determines to what extent he is unconsciously sticking to methods of solving problems which have previously proved applicable but which are unsuitable or impossible to use with a new type of problem. If a person is under stress his tendency to use methods of problem solution which have worked before increases. This means that people unfamiliar with tests who have been brought up in such a way that it is extraordinarily important for them to do well in the test often get lower scores than other individuals who are under less stress in a test situation.

Persistence

For many problems, and especially for problems in untimed tests, the longer an individual puts his mind to them, the greater chance he has of solving them correctly Persistence can not, of course, be separated from most of the other factors already mentioned.

Speed

A lot of people can be very quick at answering verbal questions but very slow at doing sums. Many people are naturally slow in general. There are slow individuals who, given sufficient time, almost always reason their way to the correct solution. Individuals of this type would get very low points in most intelligence tests because they are timed.

Definition of concepts

The concept is the most important material of the thinking process. (This applies to the thinking processes in intelligence-test problems: in non-literate and above all in non-linguistic connections what will be involved are mental images of parts of a previous milieu). It is obvious that with the problem "In what way are loud and soft alike?" (Stanford-Binet), the answer will pass only in cases where the concept "loud" has the same meaning for the testee as for the designer of the test. In order for two individuals to get the same points in intelligence tests it is not enough for them to have as much knowledge as each other, they must have the *same* conceptual definitions, "speak the same language".

The factors listed here, together with the size and quality of the data store and values cover the majority of the variations in IQ between different individuals. So it is interesting to enquire which of these factors are genetically based and which are the result of different environmental events before, during and after birth. The factors which *seem to be* most genetically based are concentration, speed and memory.

To begin with concentration, this may depend on a particular genetic disposition of varying quality in different races and individuals. But it is a fact that people's power of concentration can be changed if we change their degree of motivation and/or direction of interest in different ways. A good way of changing a man's degree of concentration is to frighten him and lead him to expect unpleasantness in the immediate future. Environmental factors are probably more significant to the variations in concentration between individuals than the basic genetic disposition.

Speed is intimately connected with physiological conditions, such as glandular activity, vision, hearing and general reaction capacity. It is affected by different environmental factors which can be seen, for instance, when comparing the general manner of speech and movement of town and country dwellers. It is quite possible that the variations in the speed with which the thought processes and reactions take place in different individuals are based to a great extent on differences in their genetic constitution. An experiment by Friescheisen-Köhler on the speed of different individuals, defined as the aggregate speed of knocking (voluntary act) and of ability to follow the rapid ticking of a metronome (perceptual act) resulted in the following figures: individuals who were unrelated varied on average by 19.5 units, non-identical twins by 15 units, siblings in general by 14.5 units and identical twins by 7.8 units.

Memory is usually divided up and regarded as two different capacities: memory span, and retentiveness over long periods. Memory span and distractibility are intimately connected. Experiments have resulted in a negative correlation of about –0.30. Moldavsky and Kaye showed in an experiment that factors affecting anxiety in the test situation also affected memory span.

Retentiveness has been shown in various studies to have a very low association with intelligence-test results. Long series of electric shocks caused retentiveness to deteriorate, as did major brain damage, regardless of the site of the damage. Most probably, retention is affected by recent events or recently learned material. Other studies have led to the conclusion that retentiveness cannot be separated from the capacity to learn data.

Probably both memory span and retentiveness are variations partly dependent on the genetic constitution of the individual. However, it is also possible that in practice motivation factors and attitudes play a greater role in the variations between individuals' memories than their different genetic disposition.

To sum up, it could be said that all the factors underlying differences in IQ between the individual who has 90 points and the individual who has 140 points are only to a very small extent genetically based and that in practice the influence of the environment is probably of greater significance to the actual difference in IQ scores.

19

CAN INTELLIGENCE BE TRAINED?

In 1946 Bernardine Schmidt published a study which aroused controversy because it provided far too positive proof that intelligence could be trained. Since one of the most important ideas underlying the concept of intelligence was that it was inherent and constant for each individual throughout his life, her results constituted a serious body-blow to the established system.

Bernardine Schmidt took over 254 schoolchildren of 13 who had been in remedial classes. The average IQ of the group was 52 when she began her three-year training period. The children were taught how to learn to read and write and study in the right way. At the same time Bernardine Schmidt adapted her instruction to the children's requirements. Only the subjects which really interested them were pursued. She was also very quick to encourage pupils as soon as they made progress.

After the three-year training period the children were followed up with tests for a further five years. Since the children were tested from time to time throughout the eight year period, it was possible to observe that their IQ not only increased markedly during the training period itself, but also continued to increase in the succeeding years. This was because the children's ability to read, write and study enabled them to make better use of their school time and gave them increased intellectual self-confidence and interest in learning.

In the last test, eight years after the start of the experiment, the average IQ of the children was 93. This is quite fantastic, considering that they were classified as morons at the beginning of the experiment.

An examination of the experiment shows that the increase in average IQ resulted to a great extent from treating the children lovingly and from timely attention to their linguistic handicaps, which were removed in the training. Moreover, for various reasons, the children were probably for the most part not organically brain-damaged, but had been neglected by their parents. Part of the increase in IQ may have resulted from the frequent testing of the children, giving them a special capacity to solve intelligence test problems. But even when these factors are taken into account, the fact remains that the leap from an IQ of 52 to 93 was enormous and that further checking of the children when they were adults and working in the community showed that they really had greatly increased their intellectual and social competence.

Where intelligence is concerned there is a rule-of-thumb which says that a low IQ tends to get lower still with age, whereas a high IQ tends to increase. This is because the IQ is also a measure of the ability to handle language. Someone who cannot handle the written language well and who finds difficulty in understanding written instructions and explanations quickly loses interest in the intellectual side of life. This further reduces his ability to pass intelligence tests, whereas the opposite applies to people whose IQ is high from the start.

Training of the capacity to pass intelligence tests should be distinguished from intellectual training as a whole. It has been found that even taking the Stanford-Binet test once increases the IQ by an average of 2.5 points.

Many experiments have been carried out on special training of the capacity to solve problems of the type which occur in intelligence tests. All have demonstrated that this special training does increase the capacity. The increase is greatest for individuals whose IQ was low to start with.

Even a little special training has produced good results in test performances. People have been trained in logical thinking in the widest sense and in semantics (the science of words and their meanings). The IQ of college students trained in this way increased significantly. Other studies involving training programmes have produced only small IQ increases and a few have produced no increase at all. It is all too easy to train and educate people of low intelligence in such a way that they find the instruction uninteresting or "above their heads." This does not mean that the psychologists who undertook the unsuccessful training experiments deliberately administered unsuitable training. What it probably does mean is that high scholarly qualifications do not always go hand in hand with teaching ability and understanding of individuals of vastly inferior intellectual capacity.

There is a good deal of data to indicate that experience and learning in babyhood and early childhood are important to the individual's future IQ. The psychologists H. M. Skeels and H. B. Dye worked with 13 infants between the ages of 7 and 30 months who lived in a home for the developmentally disturbed. Their average IQ at the beginning of the experiment was 64. The children were placed in special departments in the home, and some girls of below average intelligence (50–75) who liked small children were allowed to look after them. They spent their time with the infants, played with them, cuddled them, etc. This proved to be an exceptionally fine psychological environment for the children. After only a few months the average IQ of the infants had increased by 28 points. Since they were infants not too much significance should be ascribed to these figures as regards future intelligence, but within the given period their intelligence did increase enormously.

Experiences in the first few months of life play a major role in the IQ of adults but we do not know quite how. Probably these early experiences produce a sort of general attitude to new events and circumstances: either curiosity about their surroundings or evasive reactions to surroundings and changes in them; an introverted or extroverted attitude, anxiety or calm, etc., etc.

The experiences of the first months of life apparently remain with the individual for good, affecting not only his social relationships, choice of profession and tastes, but also his intelligence. One might expect the IQ of adopted children not to be close to their adoptive mothers' IQs, but to have been "imprinted" in one way or another by their experiences with their biological mothers in the first few months of life. It is, after all, the child's biological mother who has been its most important environmental factor in those important early months.

Skodale and Skeels' study of adopted children and IQ also showed that the adopted child's IQ was on average considerably nearer to that of its adoptive mother than that of its biological mother. *But* the correlation between the IQ of the biological mothers and their own childrens' IQ was higher than the correlation between the adoptive mother's IQ and that of their adopted children. In other words, the children's relative level of intelligence corresponded more nearly to the relative level of their biological mothers than to the relative level of their adoptive mothers, although the children's IQ points were closer to those of their adoptive mothers. (If this sounds obscure, look at Figure 34.)

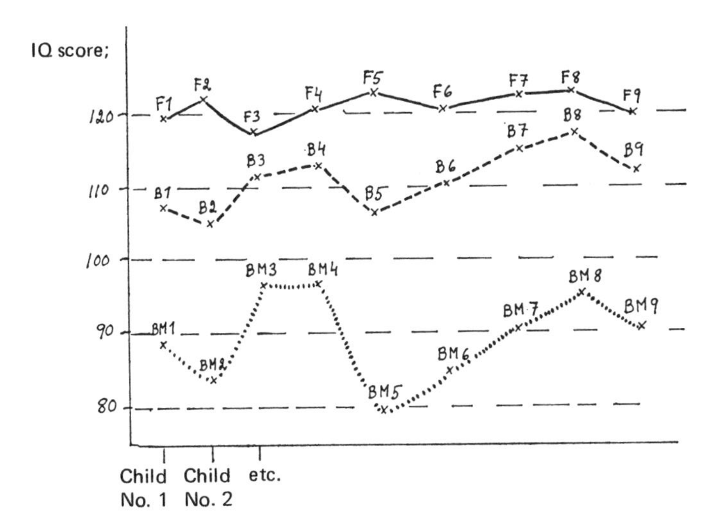

Figure 34. The relationship between the intelligence of adopted children (B), their adoptive mothers (F) and their biological mothers (BM). The IQs of the children are closer to those of their adoptive mothers than to their biological mothers. Nevertheless, the correlation between the IQ of the children and their biological mothers is higher: the line of dashes and the line of dots rise and fall in the same pattern. The uninterrupted line representing the IQs of the adoptive mothers rises and falls in a different pattern from the lines of dashes and dots.

Another study showed that it is no use at all comparing the IQ scores of adopted children and natural children. The situation of being an adopted child is completely different, psychologically speaking, from the own-child situation. Of a total of 180 adopted children, 82 had been placed in pairs in the same adoptive homes. This study showed that the correlation between the IQ of the children in each pair was 0.65, exactly the correlation which prevails between the IQs of non-identical twins who have grown up together. But the adopted children were not related to each other at all. The correlation between the IQ of the adopted children studied and that of their adopted siblings, i.e. the children of the adopted children studied and that of their adopted siblings, i.e. the children of their adoptive parents, was 0.21. (Ordinary siblings who have grown up together have a correlation of 0.52.)

These data seem to indicate that in general adoptive parents treat their own children a little differently from the way they treat their adopted children, and that these small differences in the psychological climate of the home environment are of great significance to the development of the child's intelligence.

The training of individuals in order to increase their capacity to gain high scores in intelligence tests must take account of such subtle psychological factors, which affect confidence in the capacity to think and reason. I personally think that the most important contribution is to remove the mystery as regards the relationship between symbol and reality. As soon as a person has fully appreciated that each word and each figure is a sign with a definite meaning, and no longer has the illusion that there is some mystical connection which he does not understand between the symbols and their meaning, he can begin to learn to handle them like bricks in a game. Take a little girl, for instance, who has failed to answer correctly the question: "A bird flies, a fish . . . ?" Until someone helps her to understand the test designer's rules of play for these symbols she will regard it as an absolute mystery that "A bird flies, a fish lives in water and swims" is an unacceptable answer.

"Word-magic" is a term which describes the tendency of

children and people of illiterate background to base their thinking on the idea that there is a real thing or circumstance underlying each word. Hearing is believing: legends are based on this word-magic. If there is a legend about a green dragon which belches flame, then of course somewhere in the world there is a green dragon which belches flame. When thinking based on word-magic disappears, the capacity of legends to fascinate children also disappears.

20

INTELLIGENCE TESTS AND THE SOCIAL SITUATION THEY IMPLY

The social situation in which tests are taken is interesting. Here we have a group of people to be tested. The test will show the examiner how much power of thought the testee has inherited.

All the testees are more or less tense. Those who believe themselves to be more intelligent than most of the group feel happily excited. Those who know that they are stupider than most of the others feel dully unhappy, and probably also feel an apathetic depression at the thought of their unchangeable heredity, their parents and grandparents.

When at last the test reaches the individual desks and the testees are allowed to open the envelopes, the pressure is lightened. The problems are quite easy. The first problems are always easy, in order not to discourage the worst of the bunch right from the start. All test designers know that every failure has the effect of reducing interest in and ability to solve the remaining problems.

If the context is military enlistment many of the testees will be thinking "Shall I make myself out stupider than I am? Will they see through it? What will they do if they notice?" Ideas about the length of service, the magnitude of the reward, the status of a non-commissioned officer, the job, the future and the girl-friend clash.

(Apropos nothing at all, I would like to add here that there are some psychological tests, both intelligence tests and tests of feelings, attitudes and mental deviations, which are intended to unmask anyone who is giving deceptive answers. In a number of personality tests questions are inserted which are intended to detect such pretences. But if the testee is completely consistent, and this is not too difficult when you're filling in a form, he can never be unmasked.)

In individual tests the situation is a little different. Here the testee sits face to face with an enigmatic and generally friendly man or woman who is going to assess his intelligence. The testee knows that his only chance of finding out how well or badly he is doing is to observe the examiner's expression when he has performed a task. The examiner, for his part, has been thoroughly instructed not to give away anything by his expression, but to be friendly and encouraging.

If this is to be an assessment of mental responsibility, or if the testee is from the lowest social groups in the community, he probably feels some hostility towards the man on the other side of the table, and a certain apathy, in the consciousness that he is an inferior individual. In this case his principal interest is in keeping on the right side of the examiner. He knows from the start that he will not do particularly well in the test questions.

When the problems arrive, at least one mystery is cleared up for the testee. "Well, is that all there is to it? Is this supposed to measure my intelligence? can there be anything behind it?" are common reactions.

To a member of a Western society the test situation as such is not strange, but it is strange to an individual from a pre-literate or oriental culture of a type which attaches great value to cooperation and little to competition. In the section about other races and their intelligence we find, for instance, the difficulty Australian aborigines have in understanding why the examiner and they should not *help each other* to solve the problems.

In this generally tense social situation, otherwise insignificant external circumstances will assume exaggerated importance. One study showed that students who were allowed to use desks got higher points than students who were obliged to write on surfaces joined to the chair. Another study showed significant differences in results in cases where the examiner smiled and said "Fine," "Good," and so on, in connection with the solutions to the problems.

A study of a test situation in which Goodenough's

Draw-a-Man Test was given to young schoolchildren showed the following: in its first test the class had just written a composition on "the best thing that ever happened to me." Just before the second test they had written a composition on "The worst thing that ever happened to me". The IQ scores from this second test were between 4 and 5 points below the scores from the first test.

In almost all tests it is possible to guess the answers to questions which one cannot answer in any other way. It is largely up to the examiner whether the testee, on the one hand, realizes that this possibility exists, and on the other hand, dares to make use of it. In this respect there are marked, culture-based differences between different ethnic groups. Among some Indian tribes, for instance, it is regarded as impolite to answer a question if one is not quite certain that the answer is correct.

21

THE REAL FUNCTION OF INTELLIGENCE TESTS IN SOCIETY

The most important intelligence tests in any country have several implicit functions. One is to act as a tool for the selection of individuals who will assume leading positions in the present structure of society, i.e. the psychiatrists, town-planners, doctors, civil engineers, lawyers, etc.

Intelligence tests define sound commonsense. They are used in almost all mental examinations. Sound commonsense is involved when someone seriously questions the prevailing structure. "But you understand. It wouldn't work. What would it be like if everyone . . . ? Your own sound commonsense should tell you that!"

The most important intelligence tests are verbal. Therefore the test also functions as a linguistic standard. The meaning of words is changing all the time. Individuals who are accustomed to the language of a particular social group, a sub-group for instance, will not have their definitions accepted and those who want to get on in a society have to resign themselves to learning the definitions of the leading group (that of the test designers).

Intelligence tests also imply a definition of the highest human capacities. Hence thinking based on Western logic must be the highest human capacity. It is taken for granted that there is only one correct way of thinking and that the basic Western symbol conventions are the only ones which exist. Fundamental to Western logic is the sharp dividing line between the concepts described by one symbol and the concepts described by the symbol which negates the first symbol. In logical language: *'A' is not 'not-A'*. This fundamental symbol convention is reflected in the tendency to draw sharp boundary lines between mental activity and physical events. The oak tree in the meadow is *either* a hallucination *or* something which exists in th physical universe. The fact that the oak tree in th meadow together with a number of other circumstance are *the experience of one or more people* is less significan than the conceptual category in which it should b classified.

Western logic has a tendency to disregard the fact tha all assertions, ideas, concepts and definitions of physica objects and events depend on mental activity. Huma consciousness is assumed to be a kind of mirror, whic reflects an objective independent physical reality. Thi means that two things are ignored:

1 It is not certain, and indeed it is impossible t prove, that there is a physical world which exist independently of human consciousness.
2. If there is such a world, our reflection of it, wha we see, hear, etc., depends on the conceptual syster with which we are indoctrinated. All our scientificall tenable assertions about this possibly existin physical world are fundamentally based on juxtaposition of our own and other individuals sensory impressions, which are "filtered" through th conceptual system.

Since intelligence tests tacitly assume that the Wester way of thinking is the only right one, they are a kind o definition of superman.

22

INTELLIGENCE TESTS AND THE MENTAL ATTITUDES OF OTHER CULTURES

Some intelligence tests claim to measure the intelligence of people from other races and cultures in such a way that the results are comparable. The idea that the result should be comparable is connected with the idea that it is possible to obtain an exact measurement of native capacity to think which does not change appreciably during the individual's lifetime and which is independent of cultural circumstances.

Of course, the way the individual thinks depends on the concepts he has learned while growing up.

Professor Anne Anastasi observes that it is impossible to design culture-free intelligence tests, because no man lives in a cultural vacuum. She writes, "Nevertheless it is theoretically possible to construct a test which presupposes only experiences that are common to different cultures."

More recently we have stopped using the label "culture-free" about intelligence tests and have instead begun to use the term "culture-fair."

Culture-fair tests consist of non-verbal problems. Raven's Progressive Matrices (Figure 36) and STI (Figure 35) are examples of such tests. Raven's Matrices

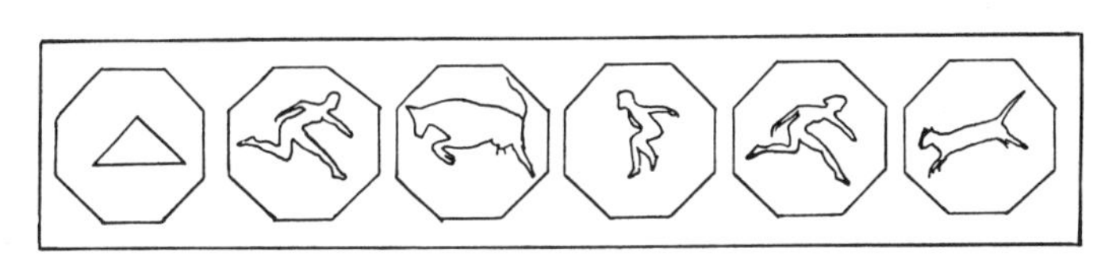

Figure 35. Part of problem from STI (see section on different types of intelligence tests). For us Westerners it is obvious that the geometrical figure on the far left must be the most important. It is the most "different", most "abstract", "first in line", etc.

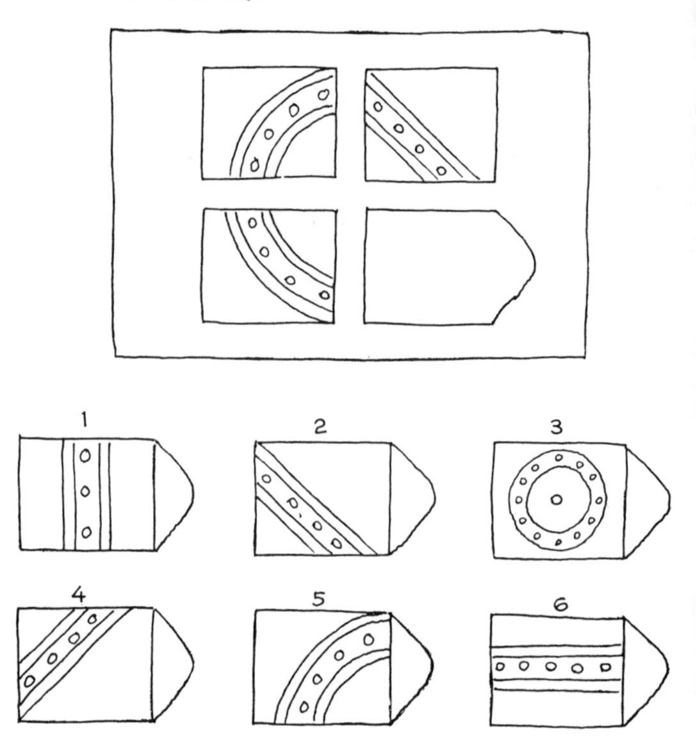

Figure 36. Problem from Raven's Matrices. If a man from a culture where symmetry is not highly valued (i.e. most illiterate cultures) is given this problem he may choose some piece other than number four to complete the pattern. This makes him unintelligent.

presuppose a knowledge of several conventions. Even the STI presupposes at least one convention, namely that the geometrically simple form must be the most important of all the forms in a series. This is obvious to us. For an African from an illiterate culture it might be more natural to ascribe the same degree of mutual importance to all the pictures, or to ascribe the greatest importance to the shape representing a cow, for instance.

This problem from STI assumes familiarity with some or all of the conventions which seem so obvious to the literate individual. For him it is natural that the shape furthest to the left or furthest to the right in a series of pictures of similar size will be the most important. Even an

Arab, reading from right to left and back to front (from our standpoint) is somewhat disadvantaged in relation to problems such as those in the STI. An Australian aboriginal has no chance at all until he has been taught how the alphabet works. But STI is still fairer than most culture-fair tests.

Raven's Matrices seem fairer still. They require a knowledge of the symmetry principle and the adoption of the convention that the symmetrical pattern is more correct than the asymmetrical. The nearest nature itself comes to absolute symmetry is the mirror image (in a pool of water, for instance). But this is not completely symmetrical. It is inside out and back to front (very well described in Martin Gardner's "The Ambidextrous Universe"). When someone from an alien culture is faced with Raven's Matrices it may occur to him to select something other than the "right" pattern. He does not know that complete symmetry is the basic idea in Raven's Matrices. To someone who has not grown up in a culture with geometry and geometrical shapes, the principle of symmetry is not apparent (see Figure 36). Moreover, Raven's Matrices require familiarity with geometrical figures in looking for the relations on which the symmetry principle is based.

The Cattell Culture-free Test includes the task shown in Figure 37. It is quite obvious that no one can solve this

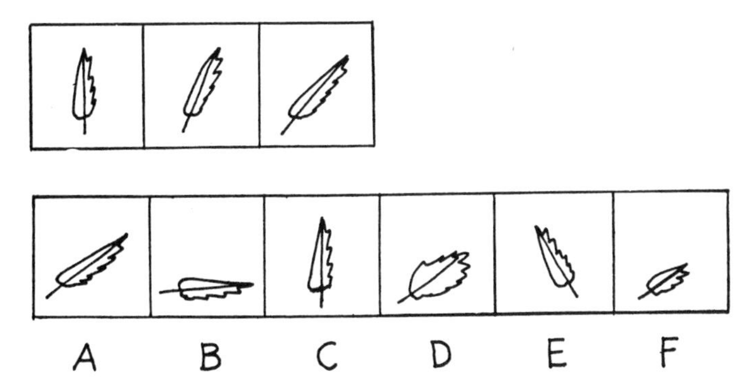

Figure 37. Problem from Cattell Culture-free Test. Choose which of the six pictures of a feather marked with a letter should follow the three pictures at the top.

problem without knowing that the three pictures at the top must be followed by one of six pictures in the lower series and that the four pictures together are to form a series. Not just any old series, but a series of a special kind, in which the moving factor has to move like the hand on a clock whose position is shown at four points in time, with exactly the same interval between them. It is quite easy to imagine a case where an individual with a highly developed capacity for thought and imagination from another culture is given this task, can see that the pictures represent feathers, can also become aware that the pictures represent the same feather in different positions. This clear-thinking individual might then reach the following conclusions: "The picture series represents a feather which is falling (from a bird) to the ground. Everything which falls towards the ground lands on it sooner or later. So the picture in which the feather is lying horizontally must be the one to complete the series" (assuming that in his culture the convention prevails that objects lying on the ground are represented by figures drawn horizontally in relation to the rest of the picture).

This reasoning is graphically demonstrated in Figure 38. Since the answer envisaged by the test designer is different, this clear-thinking individual from an alien culture would be assessed as unintelligent. Yet his answer is completely in accordance with (Western) logical thinking. From our aboriginal's point of view the expected answer to the problem is the least probable and the most illogical. And in fact of course the feather does not represent a feather but the hand on a clock. In other words, what the test designer is trying to elicit is the capacity to grasp a pictorial sequence as a regular movement divided by exactly equal intervals. This capacity is connected with the Gutenberg Galaxy and does not possess the same degree of significance in illiterate cultures. (Term "Gutenberg Galaxy" was introduced by Professor Marshall McLuhan. The "Gutenberg Galaxy" is the sum of all human sensory impressions, "filtered" through an alphabet-indoctrinated awareness.) For the "right" answer, see Figure 39.

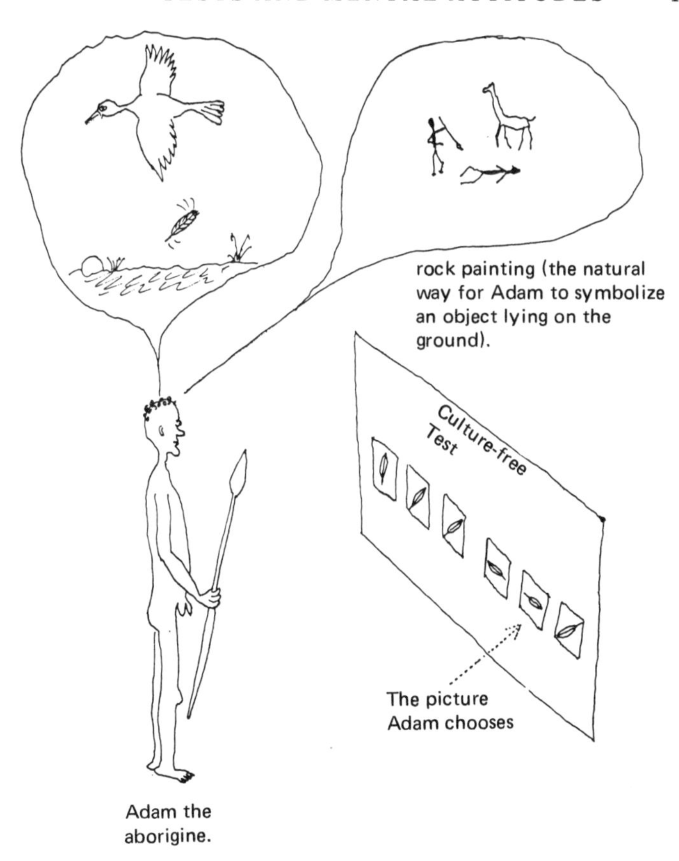

Figure 38. Aborigine showing poor mental ability in a so-called culture-free intelligence test.

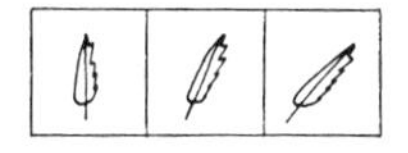

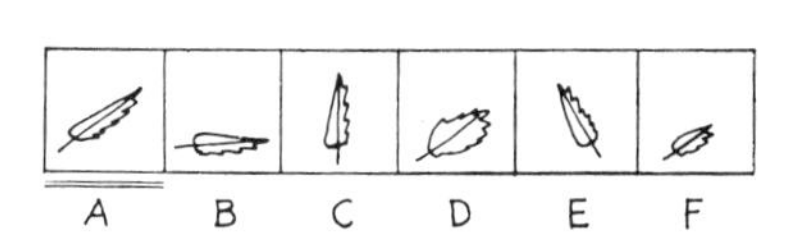

Figure 39. The "right" answer to the problem in Figure 37. The "right" feather is the one which is underlined.

23

INTELLIGENCE AND DIFFERENT RACES

Earlier in the 20th century psychologists were more convinced than they are now that intelligence was inherited. There are inherited differences in appearance and physical structure between different groups of people, of the same type as the differences between the brothers and sisters in one family. Consequently it was assumed that there would be differences in intelligence between different ethnic groups (races), corresponding to the differences in intelligence between brothers and sisters in the same family.

It was demonstrated in the USA that Negroes and Indians, for instance, had a considerably lower IQ than the whites. It was also demonstrated that white Americans descended from different nationalities, had different IQ levels. The Army Alpha test of 1917 listed the intelligence test results from large groups of individuals born in countries other than the USA in the following order:

England	Ireland
Scotland	Austria
Holland	Turkey
Germany	Greece
Denmark	Russia
Sweden	Italy
Norway	Poland
Belgium	

England, at the top, had about 15 points, while Poland, at the bottom, had about 11 points. Here the figure 14 corresponded to an IQ of 100.

The publication of this list led to the justification of those who regarded the Nordic races as superior. The

Southern Europeans had lower results than the Northern Europeans and the Germanic people had a higher intelligence than the Slavic. Those who questioned the theory of racial mental superiority thought that the variations in the average IQ were explained by the fact that immigrants to the USA originated from different social strata in their respective countries of origin. That the variations might be due in large measure to differences in degree of knowledge of the English language was something very few people considered. All the testees could speak English, after all, so the differences in the test results must reflect native capacity.

Galton believed fully and firmly in great variations in intelligence between the races. It was in order to improve the heredity of the English people that he left his fortune to the creation of an institute of eugenics in London on his death. Galton had designed an intelligence scale which applied to the whole of humanity. It had 16 steps. On the basis of his own observations and historical studies, reports from journeys of exploration and other data, he considered himself justified in stating that Negroes in general were two steps below the Englishman in mental capacity. The Englishman in turn was two steps below the average Greek in Athens in the third century before Christ, who represented the highest mental level of mankind until now.

Not many psychologists have shared this belief, in the form in which it is expressed. But many have had ideas to the same effect, and still have them. One example is Professor Henry E. Garrett, at Columbia University. In a textbook used at Swedish universities he writes, "No one is seriously going to claim that the Hottentots have shown the same capacity for the achievement of goals as the European inhabitants of South Africa, or that the Eskimos have shown the same gift for scientific discovery as modern Germans, for instance. In other words, Galton's view that there are inherent differences in intelligence between races which are widely separated on a scale made up according to what the different races have achieved seems justified. As Thorndike (Professor of Pedagogics in the USA) so aptly expresses it: 'Common observations of Africans and Europeans, for example, show that the latter are superior in intellect, initiative and self-confidence .... Two races need not be equally gifted simply because they have adapted equally well to their environment, if one race through superior initiative has created a more demanding but also a more productive environment.... Perhaps the Bushman can count all he needs to count but to put oneself in a position which requires algebra and calculus may in itself be a symptom of superiority.... The fact that a particular test seems unfair when applied to a Bushman may be a proof of his inferiority.' "

Garrett's and Thorndike's arguments are interesting as an example of how scientists of high education mix up value judgments with observed facts and arrive at conclusions which border on the ridiculous.

One of the most important published studies based on the race theory was made by Klineberg in about 1930. He tested 7 groups, with one hundred school children between ten and twelve years old in each group. The groups were taken from agricultural communities in France, Germany and Italy. The communities in which the children lived had been selected on the basis of the fact that the population there was regarded as particularly representative of a certain racial group. Only those children were selected who were good representatives of their race as regards appearance and anthropological measurements. Three races were studied: the Nordic, the Alpine and the Mediterranean.

Klineberg found significant differences in intelligence between the groups of schoolchildren when they were grouped by nationality. The German children were more intelligent than the French, who in turn were more intelligent than the Italian. But when the children were grouped according to race the differences were not obvious. Professor Anastasi writes: "Whatever differences were found, therefore, were associated with the predominantly cultural category of nationality, rather than with physical type.... Although the highest mean score is obtained by a Nordic group, the highest median (which implies that there were more 'highly intelligent' ratings) is found in a Mediterranean group.... In France the Mediterranean group is best, the Alpine intermediate and the Nordic poorest; whereas in Germany the Nordic is superior . . . and in Italy the Alpine."

In connection with this study Klineberg also tested children from Paris, Hamburg and Rome. The differences in intelligence between the urban groups and the rural groups was considerably greater than either the racial or the national differences in intelligence.

A study in New York of 400 students of different racial types gave similar results. This was not an intelligence test but a test of various personality traits, temperamental differences, etc. The results showed that the physical racial type was not related to psychological personality features to any significant extent.

Professor Garrett, quoted above, has said that there is considerable evidence that Indians with white blood are superior to full-blooded Indians in the same way as mulattos in general are superior to full-blooded Negroes. The following study may therefore be of interest:

A group of 63 highly gifted Negro children was divided into four groups according to degree of racial mixture with whites. If the racial theory were correct, one would expect that the percentage of mulattos among these highly intelligent children would be higher than the percentage of people of mixed race in the Negro population as a whole. Instead the study showed that rather more children in the highly intelligent group than might have been expected were more Negro than white (more than half their grandparents on both sides were Negro). The highest IQ rating in the group was achieved by a Negro girl who had no white blood at all. She had an IQ of 200.

Racial studies of this kind are dubious, because of the difficulty of estimating the proportion of the Negro population as a whole which is predominantly of Negro origin or white origin, respectively. Such an estimate will always involve powerful subjective elements, because the only way of estimating the origin of large groups is based on an estimation of the average skin colour.

Quite apart from the uncertainty of this type of study, the fact that among the Negro population there is an individual with an IQ of 200 means (if one subscribes to the hereditary doctrine) that there is a disposition towards hyper-intelligence in the Negro race and that the inferiority of the average intelligence of Negroes is exclusively the result of how many children the different intelligence-groups in the race produce, in relation to each other.

A résumé of more than 200 studies of Negro intelligence compared with that of whites, made by Shuey in 1958, shows that the whites had an average rating of 100 as against the Negro 85. Of course, this may result from differences in the distribution of the two ethnic groups. If one believes in the genetic doctrine of intelligence one must assume that the disposition towards an extremely good central nervous system is poorly distributed in the Negro population, or that there is a dominant disposition towards a less good nervous system which suppresses the good disposition, or one must make some other, similar assumption.

In view of what we know about the history and present social position in the community of the American Negro population the genetic theory seems foolish. Studies show, among other things, that the difference between white children and Negro children in general is least when the children are youngest, and increases with age. Psychologists working in the early 1900s interpreted this to mean that a Negro child's mental growth ended earlier than the white child's. This was assumed, because it was observed that Negroes' IQ, instead of being constant, diminished as they grew older. But as the section on intelligence and environment shows, a fall in IQ is just what we can expect in a group of people who have not the same opportunities for school attendance and intellectual occupations as other groups in the community. The concept of the IQ is in fact built up round an average value for the population which reflects its average schooling and intellectual occupations. Any group with an education and training situation inferior to the average of the population will automatically produce IQ figures which decrease with the years. And it is quite obvious that the longer one goes to school and the better the school, the higher the IQ will become. The IQ is constant only for individuals who have received an education corresponding to the average education of the population.

We have also seen that mothers from the lower classes often eat insufficiently nourishing food (cola and deep-frozen hamburgers), and that the poor nutrition during pregnancy leads to a lower IQ in the offspring. We have also seen that exceptional heat in the summer months increases the frequency of mental defectiveness. The heat is naturally greatest in the big cities and air-conditioning is still generally restricted to the upper classes. In short, there are many factors which would lead us to expect a much lower average IQ in the Negro population.

But these factors alone cannot explain such a large average difference as the one which actually exists. There is another, and probably considerably more essential factor. Intelligence tests are an expression of a long cultural tradition, a certain way of thinking and a certain way of evaluating things such as time, impulsiveness as opposed to rigidity, acceptance as opposed to independence, adherence to patterns as opposed to creativity. More over, they are in most cases based on the most specific and advanced instrument of communication in the human culture – language

Professor Marshall McLuhan at the University of Toronto in Canada has shown what an enormous difference arises in modes of thought and of appreciating one's surroundings when a society makes the transition from an illiterate, hearing-centred appreciation of reality to a vision-centred, linguistic appreciation of reality. McLuhan writes:

> "Western history was shaped for some 3,000 years by the introduction of the phonetic alphabet, a medium which depends solely on the eye for comprehension. The alphabet is a construct of fragmented bits and parts which have no semantic meaning in themselves and which must be strung together in a line, bead-like, and in a prescribed order. Its use fostered and encouraged the habit of perceiving all environment in visual and spatial terms – particularly in

> terms of a place and a time that are uniform, c,o,n,t,i,n,u,o,u,s and c-o-n-n-e-c-t-e-d. The line, the continuum . . . became the organizing principle of life. 'As we begin so shall we go on.' 'Rationality' and logic came to depend on the presentation of connected and sequential facts and concepts. . . ."

In the section on intelligence and thought in other cultures there is an example (see Fig. 37) of a problem from a culture-free test. A problem which for us Westerners seems so obviously free from cultural factors that even highly qualified psychologists have classed it as culture-free. In fact it is not at all free from cultural factors. What it presupposes is a uniform, continuous and connected space and time (and a good deal more). In fact, the series of pictures in the problem and their completion is based on a number of purely Western conventions. The feather is actually a symbol of something which behaves like the hand on a clock, or a planet in space in relation to its sun. That is, the feather is supposed to sweep through equal segments in equal intervals of time. The series of pictures is a type of Western mandala, based on the same principle as the Chinese Yin-Yang and the Buddhist wheel of life.

Any intelligence test which does not involve the manipulation of form boards (see Figure 27) requires the testee to understand the Western way of conceiving time and space as continuous and linear. Most modern tests presuppose, for instance, that the testee is familiar with the relationship of congruence.

William Ivins Jr. writes on intuitive spatial concepts: "The Greeks never mentioned among the axioms and postulates of their geometry their basic assumption of congruence and yet . . . it is among the most fundamental things in Greek geometry. . . ."

Since many tests contain problems connected with pictures, people with roots in different cultures are variously handicapped in intelligence tests. There are conventions about pictures in most countries. We can immediately see that picture (a) in Figure 40 represents a semi-circle and an angle, because we have grown up in an

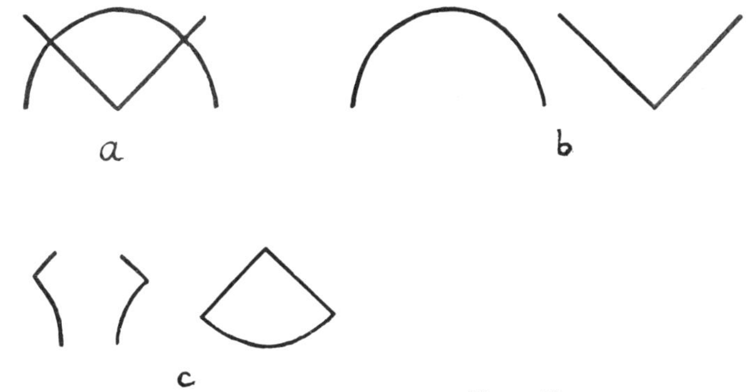

Figure 40.

environment where we are familiar with geometrical figures of this kind. For an individual from a non-geometric culture it might be more natural to divide up Figure (a) as in (c).

As far as this section is concerned, the outcome of the argument is that Negroes in the USA have not been able to read or write for more than a generation or two and that their method of thinking and appreciating their surroundings has not yet been converted to the Western way of thinking and appreciating surroundings. J. C. Carothers writes: "Whereas the Western child is early introduced to building blocks, keys in locks, water-taps and a multiplicity of items and events which can constrain him to think in terms of spatiotemporal relations and mechanical causation, the African child receives instead an education which depends much more exclusively on the spoken word and which is relatively highly charged with drama and emotion." Of course the Negroes in the USA are also brought up with keys and water-taps, but more often than white children with faulty taps, and pipes through which water sometimes comes, sometimes not, and which gurgle and knock. And often with broken locks and blocks with the corners knocked off, which are all a bit different from each other. By this I don't mean that we should pity the Negro child who does not have cubic blocks and quiet water-taps which don't drip. I mean that their environment is actually not all that Western, even now, although their forefathers have lived in the USA for 200 years and more.

They grow up in a world which is more audio-tactile than visual. There is scientific evidence for this assertion.

Many studies show that by comparison with white children, Negro children are relatively less good at tasks involving spatial appreciation and at non-verbal tests in general. And when Northern States and Southern States Negroes (who are closest to their illiterate, audio-tactile original culture) are compared, the Northern States Negroes are better at non-verbal test problems. In other words, Negroes cannot understand and manipulate geometrical objects as well as whites.

Arthur Jensen, Professor of Psychology at Berkeley University, has maintained that the Negroes in the USA have less good intelligence genes than whites, although the race has had a great many "white" (i.e. better) intelligence genes introduced into its genetic inheritance in the course of the centuries. So the Negroes in Africa should be lower down the intelligence scale than American Negroes. Professor Eysenck in England and his colleague Jencks in the USA indulge in the same ideas, even if they express them more cautiously. I personally cannot share the views of Jensen, Jencks, Eysenck and others. This inferiority, which they claim is genetically based, of Negro as against white American intelligence, is about half as great as the inferiority of white American agricultural workers to white American office workers, according to the AGCT tests from the Second World War. According to the same study, it is about as great as the average difference between white manual workers and white shop assistants. Probably the causes of this inferior IQ are quite different, and unconnected with a poorer quality of brain. On the one hand, natural, genetically-based temperamental differences probably lead to different interests and motivations from those which are typical of white Americans, and on the other hand the lower score is the result of socio-economic and cultural differences.

In this connection it might be interesting to examine the way in which culturally intact people have experienced the world. There are about 40 cases described in the literature of people who have grown up among animals, or isolated in other ways. Some have grown up among wolves, others among other animals. Most of these people have not been able to learn human speech and have retained wild and animal behaviour until their very early deaths. In general, they have shown no sexual interests, reinforcing the animal experiments which have led to the conclusion that sexual behaviour is learned early and at least for the most part not genetically based. Moreover they have always moved about on all fours in cases where they have grown up among four-footed animals. One wild girl lived on frogs and fish which she caught by swimming under water and catching them with her hands. But this is in parenthesis.

The two best documented cases are the wild boy from Aveyron and Kaspar Hauser. According to some authors Kaspar Hauser and his story are humbug. The wild boy from Aveyron was caught by three Frenchmen in the late eighteenth century. He was then about 11 years old. Kaspar Hauser was supposed to be of royal descent and to have lived shut up in a dark cell from some time in early childhood. He had been fed, but the food had been pushed into the cell in such a way that he had never been in contact with a human being. At the age of 17 he was released and was found walking along a street in Nuremberg in 1828. He could not talk, but uttered a few meaningless phrases from time to time.

Both these people showed extreme peculiarities in their appreciation of space. Neither the wild boy from Aveyron nor Kaspar Hauser, for instance, could at first distinguish between pictures of objects and real objects. It should be noted that the wild boy from Aveyron had obviously procured his food from early childhood by hunting or fishing or gathering plants, so the lack of ability to distinguish between pictures of objects and real objects was evidently no handicap out in the wild.

When Kaspar Hauser "looked out" of a window he could not see anything outside for the first few months. He saw a pattern of colours and splashes, more or less as one sees a picture. Neither Kaspar Hauser nor the wild boy from Aveyron could judge the size of objects when they were placed at varying distances from them, according to those who had charge of them. But one may wonder if these researchers confused the inability to express in words the relationships between objects and distances in a three-dimensional space with the inability to react with correct judgments of distance and size. How could a wild boy from Aveyron have fed himself without having a well-developed ability to judge distances and the sizes of the animals and plants in his environment?

J. B. Deregovsky tested the ability of Africans to see two-dimensional figures three-dimensionally. He had two groups of subjects. One group consisted of African domestic servants, who according to the hypothesis should have been able to see figures three-dimensionally because they were often in contact with European journals and papers. The second group consisted of schoolchildren who lived, outside school, in a completely African environment. So the schoolchildren should have had a greater tendency to see pictures two-dimensionally.

First the testee was given a picture of a hunter shooting an arrow at an antelope, while an elephant on a rise behind the antelope looks on. The picture is clearly three-dimensional to a Westerner. A road winds into the distance, growing narrower and narrower. Part of the antelope and hunter shield part of the elephant, and the elephant is much smaller than the antelope and the hunter.

According to Professor McLuhan a "fixed point of view" from which one sees something does not arise until a human being is literate. An observer is unthinkable to an illiterate person. Only participants are conceivable. Illiterate people are hearing-centred, that is, their reality does not consist of different viewpoints from which different aspects of something can be seen, but of different games in which they themselves are involved. This is the reason why an illiterate African Negro cannot "see" a *picture* in the same way as a white man. He "experiences" a *course of action.* No one can tell what is behind and in front and to the side, because for the African the picture does not have a point from which it is seen (see Figure 41). A picture of an Eskimo hunting situation shows the hunter on the ice, and at the same time what is behind the hunter and in front of him and underneath him (under the ice). In these pictures by illiterate artists one cannot assume any exact spatial and temporal relations from the lines of the picture

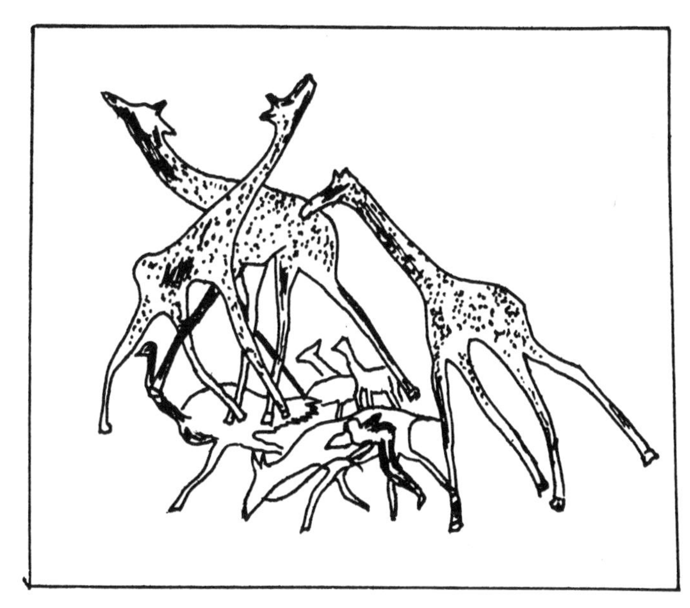

Figure 41. Rock painting from the Sahara, about 3,500 years old. In the picture a hunter is hiding "behind" an animal and stalking game with his arrow in his right hand: ostrich, antelope, giraffe. No in-front-behind relationship can be inferred from the picture. See the drawing of the giraffe's leg.

because they are not drawn from a definite observation point, as if a camera had been standing in a particular place and snapped the scene. Instead, the picture is made in such a way that what all those involved (including the chronicler) experience is mixed up and simultaneously present in the picture. The relative sizes of the figures in the picture, which are interpreted by a Westerner as meaning that objects lie at various distances from a given observer, denote instead the relative importance of the figures to the sequence of action described, or some other relationship. The pre-literate man is a listener to picture stories, the Westerner is an observer of a scene.

Deregovsky expected the domestic servant group to see the picture of the huntsman, the antelope and the elephant three-dimensionally more than the schoolchildren did, because they had had more contact with Western pictures than the schoolchildren. However, our discussion above

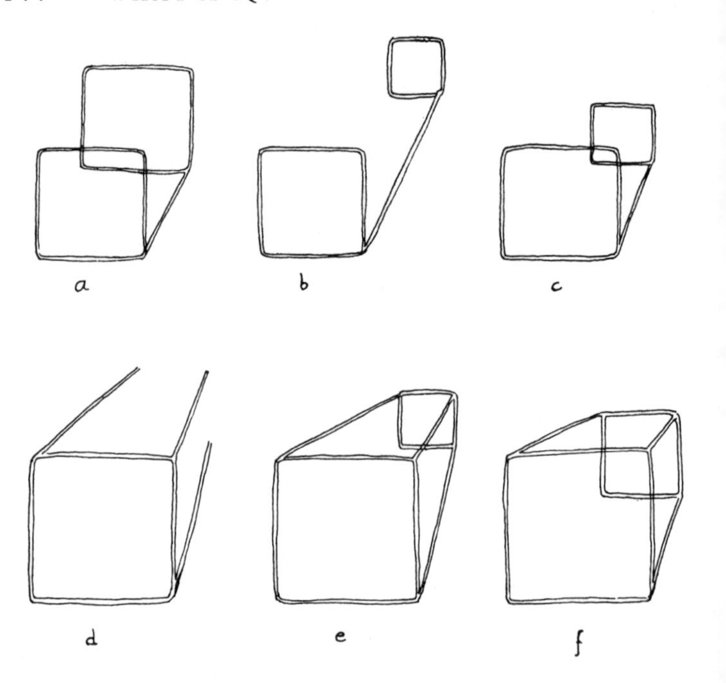

Figure 42. These are the pictures which were shown to Africans when they were given the problem: "Make something with these sticks and this modelling clay which looks like the things in this picture."

might lead us to expect the schoolchildren, who were literate, to "understand" the three-dimensionality of the picture to a greater extent. And that was how it turned out, which supports McLuhan's analysis of the way pre-literate people appreciate pictures.

The next test given to Africans consisted of pictures like the ones in Figure 42. With the pictures they were given sticks and models so that they could make both two and three-dimensional figures. The problem was: "Make a figure which looks like the one in the picture". No suggestions were made as to how the figure should be made.

If McLuhan's analysis is right one would also expect the Africans to produce mainly two-dimensional models, the schoolchildren making three-dimensional models to a greater extent than the illiterate domestic servants. This

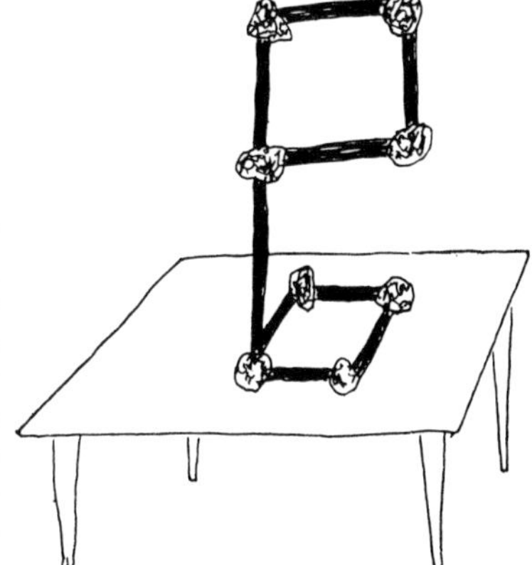

Figure 43. This model is supposed to represent (a) in Figure 42. It was made by one of the African domestic servants and indicates a rudimentary appreciation of lines in a picture in accordance with Western conventions. They are simultaneously two and three-dimensional.

proved true. The domestic servants made two-dimensional models to a greater extent, but they were often extraordinary from the Western standpoint. The model of picture (a) in Figure 42 was, in a good many cases, *semi*-three-dimensional. It consisted of a square lying on the ground with a rod sticking up and the other square placed on the upright rod like a flag on a flagstaff (see Figure 43). This means that the domestic servants had assimilated part of the Western norms of spatial appreciation and pictorial conventions and were in the process of leaving the hearing-centred outlook and adopting a "point of view", seeing the figures as three-dimensional, where the lines in the picture expressed definite spatial relationships by their position in relation to each other.

From the above discussion there is reason to suppose that individuals from ethnic groups who have a long tradition of literature and literate indoctrination behind them would achieve higher results on average than other groups in ordinary intelligence tests. The Jews are such a people. (By "people" and "group" I mean a group of people who have common traditions which distinguish their way of thinking and perceiving their environment from that of other groups of people.)

Professor Terman found that in his group of highly

gifted people there were twice as many Jews as one might have expected on the basis of the percentage of Jews in the Californian population. Moreover, intelligence tests given to the children of immigrants in New York and elsewhere have resulted in the following order of IQ rating: Polish Jews first, followed by Swedes, Englishmen, Russian Jews, Germans, etc. At the bottom of the list came Southern European children, with an average IQ of 85.

A number of studies indicate that Jewish children are particularly superior in verbal tasks. The relative Jewish superiority in intelligence tests may, of course, be based on factors other than their upbringing in a literate culture with a tradition of great respect for written language and book learning. But it is quite evident that their tradition of literacy is basically responsible for Jewish superiority in intellectual fields.

According to the same argument, American Indians might be expected to have a low IQ and different Indian tribes ought to have a higher or lower average IQ, depending on the extent to which they have assimilated Western culture. This is in fact the case. Most of the many intelligence tests made with Indians in the United States have given average values somewhat below the average for Negroes; that is, an IQ of about 80. Indian tribes which have adopted the Western life-style to a greater extent than others have had a higher average. The Osage Indians, who had the good fortune to strike oil on their reservation and therefore have a high standard of living and a higher estimation of Western values have, according to one study, an average IQ rather higher than the average for whites. Another study showed that adopted Indian children who grew up in decent white homes had an average IQ of 102. The fact that cultural factors dominate the results of intelligence tests is shown very clearly by M. E. Sparling's findings with a group of Indian children who were almost imbecile according to Stanford-Binet, but who had an IQ of 108 in the Porteus Maze Test.

One of the differences in intelligence between Indians and whites which is often referred to is the Indians' limited understanding of the importance of speed in solving a problem. Klineberg, quoted above, made a study where he compared Indians, whites and Negroes from an urban environment and a rural environment in an intelligence test which involved, among other things, timed problems. In all these problems the whites were superior. But when the rural group and the urban group of each race were compared it was found that the town dwellers were the quickest in all cases. The differences between the races seemed, in other words, to be entirely culture-based.

Differences in motivation probably play a major role in IQ variations in different ethnic groups. The tasks in intelligence tests are similar to school tasks. A white middle-school pupil feels at home in the test situation and is generally familiar with the competitive and prestige-seeking attitude which gives a good motivation to whole-hearted application. Individuals from the lower class know in general that they will have very little chance of doing well in the school situation and of having access to higher education, and are therefore generally quite uninterested either in intelligence or in tests. Probably many of them do not even take the trouble to grasp the problems but simply give up if they have not understood them at first glance.

It is also true that individuals from the lower classes of the community readily adopt an attitude of stupidity or lethargy as a defence mechanism against demands and responsibilities which the authorities might impose. This attitude, once it is thoroughly inculcated, naturally affects every sphere of life. The American psychologist F. Brown has concluded from his studies of white children and Negro children at pre-school age that there is a social pressure on the lower groups which inhibits the development of the ability to verbalize. Inarticulacy can have the effect of reducing hostility from the dominant social group and may be cultivated deliberately at first, but later unconsciously, as a measure of discretion with "survival value."

The law of the self-fulfilling prophecy also applies to intelligence. A little boy who, either because of the group he belongs to or for personal reasons, acquires the idea that he is of inferior intelligence will be inferior. All the more so since this idea of intellectual inferiority is linked in an absolutely diabolical construct with the doctrine of inheritance, which makes inferiority appear to be basically unalterable. Intellectual inferiority can therefore obviously not be compensated by training in the same way as physical inferiority or clumsiness in sports. Probably many children from large families have grasped early on the idea that they have a practical gift, while one of their brothers or sisters is the most intelligent member of the family. These intellectual inferiority complexes have a tendency to be reinforced with time. Little Billy didn't learn to read until he started school, whereas his intelligent brother Frank could read at six. Little Billy didn't dare to answer his teachers questions until he was quite certain that his answer was right, because the teacher might realize that he was not very intelligent. Because little Billy doesn't answer as often as the other children the teacher realizes that little Billy doesn't find things easy. So she gives little Billy easier questions than the other children. So little Billy realizes that the teacher has noticed he is not as intelligent as the others. Consequently little Billy is not as interested in the lessons as the other children, but sits and thinks about other, more amusing things. So little Billy doesn't hear what the teacher is saying. So he can't answer the next question, even if it is easy. So the teacher thinks that little Billy is really badly off in the intelligence stakes, and so on and so on, until little Billy has reached a position where his limited intellectual resources are not a handicap – on the conveyor belt in a factory, for instance, while little Billy's brother Frank is on the Board of Directors.

The effect of the law of the self-fulfilling prophecy is reinforced because society as a whole is indoctrinated with the idea of the biological, hereditary nature of intelligence. Those who are intellectually handicapped at an early age are therefore not given specially careful training to enable them to catch up with the others, but are set aside in special departments (remedial classes, special schools) with the stamp "nothing to be done, unfortunately," "you can't help heredity," etc., etc.

The examiner's expectations regarding performance (the Rosenthal effect) are also of some importance to the test results. In 1936 H. G. Canady published a report on an experiment in which the Stanford-Binet test was given to groups of white children and Negro children by examiners from both ethnic groups. The groups were split up so that half the children were tested by a Negro examiner first and half by a white examiner first. Then they were tested again by an examiner of the other race. Both for the white children and for the Negro children, the average IQ was six points higher when the group was tested by an individual of the same ethnic group.

I showed earlier on that cultural differences are of decisive importance to the variations in achievement in different types of tests between different ethnic groups. M. A. Straus studied students from Ceylon in the 1950s, using both verbal and performance type intelligence tests. The students were considerably superior to American students in the verbal part of the test but greatly inferior in the performance test. Straus himself ascribed these differences to the culture of Ceylon, in which manual work is poorly regarded and there is prestige in verbal studies. Another possible reason for the differences was the Sinhalese educational system, which emphasizes learning by heart and abstract concepts.

In the Goodenough Draw-a-Man Test Hopi Indian boys had an average IQ of 116, whereas the girls had a more normal IQ of 99.5, according to one study. In the Hopi culture the graphic arts are a traditionally masculine occupation.

In a study of Scandinavian children and Jewish children at the pre-school stage using Stanford-Binet, it was found that the Jewish children were superior as regards general information and verbal ability, while the Scandinavian children were superior in problems which required spatial orientation and muscular coordination.

Individuals with non-literate traditions in their cultural background should find it more difficult to appreciate and grasp linear and continuous relations because they are not as accustomed as others to perceiving reality in these dimensions. Several studies have shown that this is so. One psychologist gave 50 white and 50 Negro students some training in visual appreciation and definition of length. The white students were clearly superior at the beginning of the training but their superiority diminished with time. The ability of the Negro students was increased to a greater extent by training than that of the whites.

In the 1950s J. H. Boger carried out an experiment with white and Negro pre-school children. They were given one group intelligence test in January and another in May. Meanwhile, half the white children and half the Negro children were given practice in the art of solving problems concerning spatial appreciation, pictorial appreciation and the ability to distinguish between different symbols. In the test in May, as expected, all the trained children, whites as well as Negroes, showed an increase in IQ as against their untrained fellows. The IQ of the Negro children had increased the most. The increase in the intelligence of the Negro children was greatest in the non-verbal problems. When another test was made in October of the same year it was found that the improvement was permanent.

Australian aborigines are the people most different from the white races as regards characteristics such as brain weight, etc. This race, the most inferior and ape-like from many anthropological viewpoints, has cultural traditions which present the intelligence tester with great difficulties. The psychologist S. D. Porteus (the one who designed the Porteus Maze Test, see Fig. 21) has spent a great deal of time on tests for various non-European groups. He writes: ". . . the aborigine is used to concerted thinking. Not only is every problem in tribal life debated and settled by the council of elders, but it is always discussed until a unanimous decision is reached. On many occasions the subject of a test was evidently extremely puzzled by the fact that I would render him no assistance especially when . . . I was testing some men who were reputedly my tribal brothers (white men). This was a matter which caused considerable delay as, again and again, the subject would pause for approval or assistance in the task."

Even from the Western viewpoint the aboriginals' attitude here is actually a sign of superior thinking capacity. Experiments at Yale University have shown that problems of a different nature, including problems of the type considered to measure the degree of logical ability in Stanford-Binet, are solved more quickly and with a lower percentage of wrong answers by individuals in a group (three individuals per group in the experiment) than by single individuals. James H. Davis writes: "On average the groups produced the answers faster than those who were working individually and at the end of the first experiment had produced a high proportion of correct answers."

The fact that the most effective use of human intellectual capacity calls for group activity is exploited in the advertising industry. Big advertising bureaus often muster the staff concerned with creative activities for so-called brain-storming sessions, which means that all the people in the group turn their minds towards a definite common goal. Anyone there can throw out any idea he likes and is encouraged to do so. No criticism is allowed, at least in the first phases of the brain-storming. Since all the ideas and impulses are expressed directly and without censorship for the others in the group to hear, they in turn receive impulses which when expressed may prove to be a useful adjunct to an idea conceived by someone else.

Another result of Porteus' studies of the mental capacity of aborigines is worth noting. Porteus had been impressed by their tracking ability. He therefore constructed a test consisting of photographs of footprints, the task being to match the pictures which belonged to each other, i.e. the two prints made by the same individual. In this test the Australians achieved approximately the same results as 120 white high school children from Hawaii. Porteus comments, "Allowing for their unfamiliarity with photographs we may say, then, that with test material with which they are familiar the aborigines' ability to discriminate form and spatial relationships is at least equal to that of whites of high school standards of education and of better than average social standing."

Tests using the Porteus Maze Test have shown the IQ of the aborigines studied to be about the same as that of illiterate Chinese and Japanese.

In contrast to Negroes and Indians, Chinese and Japanese in the USA achieve about the same level as whites in their test performance. This may be because Chinese and Japanese from the lower social groups have not emigrated to the USA to any great extent. Otherwise one might have expected them to have a lower IQ than native white Americans because of differences in linguistic and cultural background. On the other hand, the most essential elements in the culture as far as test performance is concerned are on hand in China and Japan: the alphabet, the visual adjustment given by continuous linear perception categories, the high valuation of book learning and respect for the norms of the established society.

Several studies have shown that IQ variations between different ethnic groups in the USA are associated with the social selection which influences their emigration to the USA. Rose Franzblau gave non-verbal intelligence tests in 1930 to a group of Danish and a group of Italian schoolgirls in New York. She found the Danish girls to be considerably more intelligent. The next step in the study consisted of testing two representative groups of schoolgirls in Copenhagen and Rome respectively. The result of these tests showed no average variation in IQ.

The Danish psychologist Dr Kaj Spelling based his doctoral thesis on material from comprehensive tests of many types on schoolchildren in Malaysia. The population there consists of three large, relatively well-defined groups: Malay, Chinese and Indians. Spelling found average variations which were small but significant: the Chinese children had an IQ of 107, the Malays 98 and the Indians 96 on average. The children to some extent represented a group which was intellectually superior to the average population, because they were all taken from the middle school. When Spelling tested Danish schoolchildren at the corresponding level he found their average IQ to be 106.

Spelling also carried out detailed studies of the situation of the various ethnic groups in relation to their environment. He studied their language, educational circumstances, social grouping, economic status, nutritional and physiological conditions, extent of disease, etc. As an example of differences in social environment between the ethnic groups we may mention that in the Malayan group's language there is no term for "parallel." The effect of this on Malayan thinking, perception of the environment and ability to solve non-verbal problems in intelligence tests is obvious. The word for "back" (in the sense of coming back) does not exist in the language of the Malayan group either. On the other hand, there were terms in the Malay language which had no equivalents in European languages, but neither did they have any equivalents in the intelligence tests intended to measure mental capacity.

Kaj Spelling concluded from these thorough investigations, "Variations in IQ between different racial groups do exist. Variations in the social conditions of different groups also exist, which makes it probable, but does not prove, that the variations in IQ are culturally based."

P. H. DuBois carried out an interesting experiment in the late 1930s. He standardized a Draw-a-Horse Test along the same lines as Goodenough's standardized Draw-a-Man Test, but with the difference that the standardization group consisted of Indian children from a reservation in the American South. In terms of the usual validity criteria for intelligence tests, this test was more valid for the Indian group than the Draw-a-Man Test. The Draw-a-Man Test and the Draw-a-Horse Test were then given to both white and Indian children. The whites were superior to the Indians in the Draw-a-Man Test but the opposite was true of the Draw-a-Horse Test. On the basis of the Draw-a-Horse Test the average IQ of the white eleven-year-olds tested would have been 74.

24

INTELLIGENCE AND HEREDITY

The idea that heredity determines a person's intelligence is very ancient. Plato was one of the first eugenicists. His ideas on the improvement of the race were more advanced than Hitler's.

In his "Republic" Plato describes his ideal society. The people would be divided into three strictly separate classes: the *philosophers or guardians*, who would govern and direct; the *warriors*; and the *farmers and labourers*, whose principal function would be to provide the other two classes with food and other necessities. The rulers would create a myth which the other two classes of society would be induced to believe in. The myth would illustrate that in the beginning people were created in three categories. The first, the golden people, were people of the ruling class, the next, the silver people, were those of the warrior class. The third were the bronze people, those whose only function was to work for others and who lacked both courage and the capacity to think.

When the women in the state were aged between 20 and 40 the rulers would bring them together at suitable intervals with men between 25 and 55. The offspring resulting from such encounters would be taken care of by the state and brought up without knowing who their mothers were. At the ages of under 20 for women and over 55 for men intercourse could take place freely. Children of such unions should either be aborted at the foetal stage or exposed and left to die in places specially set aside for that purpose.

Family life and sentimental ideas such as love of an individual would not be entertained. In exceptional cases bad offspring from the ruling class should be moved down to the next class or a specially good product of a lower class should be promoted to a higher class.

The ruling class and the warrior class would be encouraged to breed more often than the lower class, to ensure a supply of superior individuals.

By comparison with Plato's, the eugenic visions of little corporal Hitler were touchingly humane. Hitler assumed that the Germans, that is the tall, blond and blue-eyed people, were superior to others regardless of social group. Therefore the people who showed the purest Germanic racial features should have the best opportunity for reproducing themselves. Hitler instituted a number of "love temples" on the Baltic coast, luxurious marble villas with Greek columns. To them, young blonde and blue-eyed German Girls were invited to spend week-ends and longer periods with blond, blue-eyed young SS and SA officers. The results of Hitler's eugenic experiments have not been discussed in the psychological literature. There are good grounds for assuming that the reason for this is that the children produced in this way proved to be roughly the same as anyone else, as far as intelligence was concerned.

Galton set the tone as regards the question of heredity and environment in relation to psychology. He could not – or would not – admit that the environment might have quite a powerful influence on people's way of thinking and behaving. All the variations in mental capacity between individuals were genetically based. After Galton and Binet, this idea of the exclusive importance of biological heredity gradually changed. Professor Louis P. Thorpe discusses the changing attitude in "Child Psychology and Development." There is now good evidence, he feels, to show that the traditional attitude towards the inheritance of mental abilities is beginning to change. In 1914 a study was made of 300 individuals of low intelligence at the Vineland Training School for Mental Defectives. The study indicated that 77% of the subjects had probably inherited their mental defects. Twenty years later, in 1934, the scientific head of the Vineland school at that time reported that according to his data 30% of the individuals at his institution were hereditary cases, 30% were secondary cases (damage after birth, infection and endocrinal disease) while 40% were of unknown or uncertain origin.

A number of anthropologists interpret the fact that blood relationships have been controlled among many peoples as a sign that the importance of biological heredity was recognized. Among the Pygmies in Africa and the Australian aborigines marriages between close relatives were formerly forbidden. Instruction on sexual relationships was included in the lengthy education of the aborigines to full clan membership (it took several years). The family of the aborigine boy belonged to a particular totem group, symbolized by an animal or a plant, e.g. the kangaroo totem. The totem group consisted of several closely related families. Closely related totem groups belonged to particular matrimonial classes, from which a wife could be taken. Reproduction was considerably more controlled than in our culture.

Is this an example of aboriginal superstition, or is it an expression of their superior knowledge of the genetic code? In our country we restrict ourselves to forbidding marriage between siblings. In Australia, on the other hand, the aborigines conducted an active genetic policy. They determined the hereditary groups between which new individuals should be conceived.

It is obvious that intelligence depends on what is passed down from an individual's ancestors. The point at issue is which sort of heredity do we mean – genetic heredity, or social heredity? As we have said, in the last 75 years there has been a shift from belief in the absolute dominance of genetic heredity to a growing belief in the importance of other factors. But this shift has proceeded at different speeds in different countries. British psychologists are generally more conservative than American psychologists and consider that the IQ variations in a population are for the most part genetically based. Galton's last surviving pupil, the grand old man of British psychology Professor Cyril Burt, claimed that 77% of the variations in intelligence are genetically based. American psychologists are more reticent about the percentages they employ to express their convictions, but seem to incline towards a fifty-fifty basis. And at present it is generally accepted that the effects of genetic heredity and the environment cannot be separated from each other, which is true and that genetic heredity sets a limit on the individual's intelligence potential, which means that it is pointless to train and educate individuals who have inherited a poor intelligence from their ancestors. When they have reached their "intelligence limit," further schooling is useless.

This section will be devoted to studying data relevant to the hereditary environmental arguments. The first thing to be said here is that the percentage of mental defectives among the relations of mental defectives is greater than in the normal population. An English committee on sterilization set up in the 1930s reported on the offspring of 3,700 individuals classified as mental defectives. Of some 9,000 children, 1,000 were classified as mental defectives and 2,000 had died.

In themselves, these figures do not indicate whether or not the causes of these facts were genetic. Infant mortality of this level would have been regarded as quite normal a hundred years ago, even in the highest social groups. And the high percentage of the offspring who were classified as mentally defective might, in view of our knowledge of the inability of mentally defective mothers to avoid accidents, eat the right food, etc., have been just the same if each mentally defective mother had had a newly conceived foetus with an excellent heredity implanted in her womb instead of conceiving her child herself.

Some of the data in the study indicate that it is right to apply a social interpretation to the circumstances surrounding the children of mental defectives. The offspring were divided up according to age – under and over 13. Twice as many over-13s as under-13s were classified as mentally defective. The probable reason for this is that the older the children became, the more powerful were the effects of an inadequate pre-natal and post-natal environment. It has been demonstrated in many connections that children who have had a poor early environment become progressively more backward as they grow older. There is no currently known hereditary reason for the earlier children in a family to inherit defective genes to a greater extent than the others.

It should be noted that at present it is not possible to distinguish between mental defects which have arisen as the result of unsuccessful attempts at abortion, unsuitable food during pregnancy, endocrinal disturbance in the mother, infectious diseases in the mother, injuries in labour, unintentional mishandling in infancy, etc., and mental defects caused by defective genes in the newly fertilized egg.

However, there is a small percentage of all mental defectives of whom it can be said with adequate scientific certainty that they are defective owing to poor inherited genes. This applies to between 3 and 10% of all mental defectives. At the same time, the mentally defective constitute, at most, 0.1% of the entire population. Logically speaking, it is inappropriate to use what we know about this group in order to support theories about the remaining 99.9% of a normal population and their IQs, including the association between the IQ and ancestral genes.

Galton himself trod a different road. He took the highly intelligent as evidence for his theory of heredity. It has been shown in an earlier section that there is a strong association between the IQ and the early infant environment, parental attitudes, parental educational level and the individual's schooling. It is in fact so strong that more reliable prophecies can be made when a child is 10 about his IQ at 20 by collecting data on the parents' social position than by using data from an intelligence test on the 10 year-old. Studies of highly intelligent individuals' kinship patterns are therefore worthless, from the scientific standpoint, as evidence for Galton's theories.

One may therefore wonder on what data well-known psychologists such as Cyril Burt, H. J. Eysenck, Arthur Jensen and others base their theory that the variations in intelligence in a population are predominantly genetically based.

For the time being, monozygotic ("identical") twins provide the only satisfactory evidence for a Galtonesque hereditary theory of intelligence. According to a traditional idea held by science for at least 50 years, monozygotic twins have an exactly identical heredity. (In the past the difference between identical and non-identical twins was not understood.) If there were some way in which we could procure a number of newly fertilized eggs of the identical-twin type, and if we could wait until every one of them had divided into two separate eggs, and if we could then place them in the wombs of different women from different social environments and if we were to test, at the age of 20, the individuals who emerged from this process, we would then obtain a clear decision as to the significance of genetic heredity to the variations in IQ.

Since such an experiment is impossible, the research into twins has concentrated on studies of monozygotic twins who have been separated a few years after birth, adopted by different families and grown up in different environments. This means that we cannot obtain any precise answer to the question of heredity versus environment. The events in the womb, the circumstances of the birth itself and the first year of life will be very similar for monozygotic twins. Moreover their position as adopted children will be a common environmental factor which will follow them even after their separation. This factor is extremely important in this connection. The IQ correlation between wholly unrelated adopted children in the same home, according to one study, has been shown to be 0.65, whereas the correlation between the IQs of the adopted children and the children of the adoptive mother is 0.21. In other words, monozygotic twins who have been adopted and grown up in different homes have still grown up in a similar psychological climate. Those who are convinced that the genetic code defines the limit of human mental capacity base their argument on twin-research data. This means that they regard the factors I have referred to as irrelevant to the IQ. They tend to a great extent to assume those things which actually require proof.

Be that as it may, the correlation between the IQs of monozygotic twins who have grown up in the same home has been shown to be about 0.89, whereas for monozygotic twins who have grown up in different homes it has been about 0.76.

Dizygotic twins are no more similar, genetically speaking, than ordinary siblings, according to the present scientific view. The correlation between the IQs of dizygotic twins who have grown up in the same home is about 0.63. The correlation between the IQs of ordinary siblings who have grown up in the same home is about 0.52. The correlation figures given here are the mean of the majority of dizygotic twin studies made in the last 50 years. According to these figures, variations in the IQs of monozygotic twins are less in all circumstances than for any other form of kinship.

The supporters of the theory that the environment principally determines the variations in the results of intelligence tests can appeal to the fact that monozygotic twins have a more similar environment than dizygotic ones. Professor Anastasi is one of the authors who points out that the monozygotic twin always has a contemporary "copy" of himself close to him, while the dizygotic twin has a contemporary brother or sister who is bigger or smaller than himself, more or less at home in his surroundings than himself, more or less successful in his attempts to influence his environment than himself. Dizygotic twins meet with more distinct reactions from their surroundings.

Dizygotic twins have also often attended different schools and been in different classes. In one of Professor Husén's twin studies, identical twins had the same education in 90% of cases, as against 84% for the non-identical twins. When we remember that the correlation between IQ and length of school attendance is usually about 0.70 for large groups, we see that the differences are considerable.

But the IQs of monozygotic twins are more similar than those of dizygotic twins, even when the monozygotic twins have grown up in different places and the dizygotic twins have grown up together. This means either that similarities in the infant environment and similarities in the adopted child climate, or similarities of a genetic nature, are more significant to the IQ than the same home environment while growing up.

The correlation figure for the IQ of monozygotic twins is about 0.89 and for dizygotic 0.63, when both categories of twins have grown up in the same home. For someone who is convinced of the great importance of the environment these figures imply that his is the correct standpoint. Of course, the opposite is true for the supporters of the heredity theory. Whether or not the figures given are to be interpreted as evidence of the importance of inheritance to intelligence therefore depends on one's view of the relative importance of the differences in environment. The overall result will be that one's prior convictions on the heredity-versus-environment question will be reinforced in one direction or the other, as long as correlation data continue to be regarded as having any value as evidence on the subject.

Correlation figures for the connection between the IQ of monozygotic and dizygotic twins cannot in fact be interpreted in the same way as correlation figures for other population groups. For various reasons, twins in general have an IQ which is about four points below the population average. At the same time the proportion of twins with very high IQs is about the same as in the normal population. This means that the twin group is more heterogeneous than other population groups. The correlation figures involve an over-estimation of the actual size of the correlation compared with corresponding figures for the normal population. Moreover, the death of one twin in the womb is more usual for monozygotic than for dizygotic twins, a factor which also affects other circumstances.

There are many interesting differences between the environment of monozygotic and dizygotic twins which affect correlation figures one way or the other. Dizygotic twins' parents are, on average, about five years older than the parents of monozygotic twins. Dizygotic twin births are commoner in the higher age groups, whereas monozygotic twin births are normally distributed in relation to the mothers' age. This means that the dizygotic twins come from an environment which is on average less intellectual, owing to the steady expansion of school education in recent decades which has benefited the younger parents to a greater extent. The fact that dizygotic twins generally have older parents also means that they generally have a number of older siblings. Their position in the family is liable to give them a lower average IQ than that of monozygotic twins.

There are more common differences in environment between monozygotic and dizygotic twins, besides those already mentioned. How these affect the correlation statistics is a complicated and difficult question. But for reasons which I shall now discuss, it is unnecessary to go into these in the heredity-versus-environment debate. Correlation data from twin studies are given in all the psychology text-books, whether for secondary school or university. By their position in the text they are made to stand as corroborative evidence of the relative importance of heredity in relation to environment. Twin research data are treated as if they were precise, and as if only their interpretation were problematic. But this is quite untrue.

When a researcher undertakes a twin study he must first collect pairs of twins. This is a straightforward matter. But then he has to decide which of the pairs of twins are monozygotic and which dizygotic. It is at this stage that circumstances arise which make correlation statistics from twin studies worthless as evidence in the case of heredity-versus-environment. At present there is only one recognized reliable way of deciding if a pair of twins is monozygotic or dizygotic. This is to transplant a piece of skin from one twin to the other. If it "takes" they are monozygotic, if it is rejected the twins are dizygotic. It is impossible for practical reasons to use this test in intelligence studies.

Many people think that twin status can be determined by studying the foetal membrane or placenta, but this will not do, because both dizygotic and monozygotic twins can have the same placenta and foetal membrane. Nor are fingerprints a reliable test.

For lack of other criteria, external similarities and dissimilarities between the twins have in practice often served as a basis for the determination of their genetic status. Newman, Freeman and Holzinger, who have carried out some of the most quoted twin studies, used the following criteria: striking similarity in 1) general appearance, 2) hair colour, 3) hair quality, 4) hair shape, 5) eye colour, 6) pigmentation of iris, 7) skin colour, 8) shape of nose, 9) shape of lips, 10) shape of jaw, 11) shape of ear,

12) dental type, including irregularities, 13) shape of hands and fingers.

In his book "Medicinsk Genetik" Dr Hans Olof Åkesson writes on the anthropological determination of twin status: "E. Essen-Möller was able to demonstrate in 1941 that a general assessment of individuals could produce almost as good a diagnosis of zygotic status as an evaluation and summation of individual anthropological characteristics. This observation has led to an attempt to diagnose twins by means of a simple questionnaire technique. A surprisingly good identification can in fact be made by putting simple questions of the type 'Were you often mistaken for each other as children because of your appearance?' or 'Do you yourselves think you're as alike as two pins?'"

Naturally, no predictions as to the proportional dependence on genetic heredity of the variations in intelligence can be based on a diagnosis like that. More recent medical studies on twins have therefore sought to decide which twins were monozygotic and which dizygotic by examining their blood groups. Of this Dr Åkesson writes, "On the basis of one blood group system an acceptable zygotic diagnosis cannot be achieved, but the certainty of the diagnosis increases rapidly if several gene markers are used. N. Juel-Nielsen, A. Nielsen and M. Hauge have shown in a Danish study that when they used eight blood group systems and two serum blood systems as well as sex data, over 98% of all dizygotic twins could be correctly diagnosed."

The figure of 98% in this Danish study does not mean that they had 100 pairs of twins, transplanted skin between them, compared them with probability calculations based on the determination of blood groups and found that two of the 100 pairs were able to accept the transplanted portion of skin although the calculations indicated that they were dizygotic. It is quite simply armchair speculation, the result of a probability calculation, and consequently no more correct than the assumptions about genes, their penetration, domination, etc., on which it is based. Many of these assumptions in turn are based on the results of former twin studies.

Even the most modern medical twin studies thus to

some extent assume what they are meant to be proving. But this is in parenthesis, and of no importance to the twin diagnosis on which intelligence tests are based. Almost all twin studies concerning IQ are in fact based on the anthropological diagnosis. Very few are wholly or partially based on blood group determinations and more precise probability factors.

The uncertainty of twin diagnoses meant that in every twin study one was faced with the choice of either excluding or retaining the uncertain twin pairs. Many twins are not sufficiently similar for one to be certain that they are monozygotic twins, but nor are they sufficiently dissimilar for one to be certain that they are dizygotic. If these uncertain pairs are included in the study, their classification in one or other category will be arbitrary.

If, on the other hand, the uncertain pairs are not included, the correlation coefficients for the IQs of monozygotic twins will come out too high. Some of the uncertain pairs which are not included are sure to be monozygotic twins, and they will probably have slightly more dissimilar intelligence results than other monozygotic twins.

The correlation coefficients for the IQs of dizygotic twins will be a little too low, since some of the uncertain twin pairs are certain to be dizygotic twins, and since they would probably have more similar intelligence test results than other, more dissimilar dizygotic twins.

Many intelligence research workers are honest or conscientious enough to refer to this problem. Professor Torsten Husén carried out massive twin studies at the end of the 1940s in connection with enlistment for military service. In his material 122 pairs of twins were classified as monozygotic, 193 as dizygotic and 64 as uncertain.

The correlation coefficients for the IQs of the dizygotic twins vary between 0.45 and 0.90 in different studies. It is an interesting fact that correlation figures for twin studies made in the USA generally show smaller differences between monozygotic and dizygotic twins than correlation figures from the UK. This may perhaps be explained on the assumption that British psychologists have an unconscious desire for their research results to confirm Francis

Galton's theories, whereas the Americans are less tradition-oriented. American researchers, according to this hypothesis, would have a tendency to place more uncertain cases in the monozygotic twin category than would the British.

The hypothesis I have just suggested is not an accusation against psychologists; it is based on the fact that no delimitation, where concrete criteria are not available, can avoid being influenced by the assumptions of the individual responsible for the delimitation. The Rosenthal effect is a fact.

Dr J. A. Fraser-Roberts writes in "An Introduction to Medical Genetics": "For a number of different reasons, such a twin group will almost certainly be defined subjectively. It will contain too many monozygotic pairs who are like each other, leading to an over-estimation of the role of genetic heredity."

The uncertainty of twin diagnosis makes the reported correlation coefficients so uncertain from the genetic standpoint that they lack authority in the case of heredity versus environment. Such correlation figures should be studied from the historical standpoint and might then be expected to shed light on the changing attitudes of science to the question of heredity and environment in the last 50 years.

But correlations from twin research provide valuable material for conclusions on the importance of the environment when comparisons are made, not between monozygotic and dizygotic twins, but within the respective twin groups. If, for instance, it should be found that the IQs of separated monozygotic and dizygotic twins had a tendency to be closer, the more time they had spent together before separation and adoption in different homes, then the similarity could be seen as a sign of the importance of the infant environment to the IQ. Since the environment of babyhood has hitherto been regarded as irrelevant to the IQ, no such study has, to my knowledge, been made. However, Newman, Freeman and Holzinger have made a famous study of 19 pairs of monozygotic twins who were separated in infancy and grew up in different places. This study is referred to in most psychology text-books and standard works and manuals of genetics. The study data include the ages at which the different pairs were separated. I have processed this data in order to obtain information as to the effect of environment in early infancy on IQ similarities between twins.

Newman, Freeman and Holzinger employed five expert, mutually independent assessors who allotted points for the differences in the educational circumstances of the twins in each pair. If one twin had been to a state school for six years and the other twin of the same pair had a university education, the pair was given a high points figure. The pair judged to have had the most dissimilar educational background also had the highest IQ variation: 24 IQ points. In general, the twin who had the more advantageous school career also had more IQ points, but there were exceptions. In one case a twin regarded by the assessors as having a small educational advantage had an IQ score 9 points lower than her twin sister. The fact that such exceptions occur can be interpreted as a sign that much subtler environmental factors than the quality of schooling are of great importance to the IQ.

One essential for this interpretation of the data was that the 19 pairs of twins should be genuinely monozygotic. The practical state of twin diagnosis assumes that this was so. The general practice, in fact, is to avoid uncertain cases and obtain exaggeratedly high correlation figures for monozygotic twins.

This study, as we have said, included the exact age in months at the time of separation and the age in years when tested. In order to prove the hypothesis that what happens to a child in infancy affects his IQ as an adult, I assumed that the twins who had spent the longest time in the same environment as infants would have more similar IQ ratings than the twins who were separated relatively soon after birth.

I divided the twin pairs into two groups (9 pairs in one group and 10 in the other) according to the length of time they had spent together before separation. In one half the twins were on average nine months old on separation, in the other twenty months. I then investigated the differences in IQ between the twins in each pair. I found that the variation in IQ between the twins was on average about 13% higher in the group who had been together longest before they were separated. My hypothesis appeared to be wrong.

However, the following facts also emerged:

1. The oldest twin pair, the ones who had lived apart in different environments for the longest time, had the widest variation in IQ (11.3 IQ points, as against 1.4 for the younger group). The twins in the older group were aged 36 on average, as against an average of 17 for the younger group.
2. The twins in the older group had the widest differences in school and educational advantages (20 units, as against 11 for the younger half).
3. The twins in the half who had been together for the longest time before separation were on average eight years older and had 3.5 units greater average differences in school and educational conditions than the twins in the half who had been separated relatively quickly after birth.

Since greater age and more differences in school and educational conditions meant greater differences in IQ, and since the twin group who had been longest together as infants before separation were older and had greater differences in schooling and education, one might expect their IQ to differ more. So my hypothesis might be right, although it did not appear so at first sight.

In order to check this, I selected the twin pairs who according to the assessors had had the most similar conditions of education and training. I then divided these pairs into two groups according to the time they had spent together as infants. I found that the group of twins who had been longest together had the least differences in IQ in each pair. The difference was about 3.4 IQ points, as against 5.6 IQ points for the others.

If these 19 pairs of twins are representative of the normal population, we can conclude that:

1. Differences in education and training play a major role as regards differences in IQ.
2. What happens in the first months of infancy affects the IQ of the adult.

In 1964, Bloom divided these 19 pairs of twins into two groups. One group consisted of 11 pairs who had been to very similar schools. The correlation between their IQ was 0.91. The second group consisted of 8 pairs with different educational backgrounds. The correlation between the IQ of each pair of twins in the group was 0.24. This is the same correlation figure as the one obtained in a previous study of ordinary brothers and sisters who had been separated and grown up in different homes for seven years.

Obviously this little group of people, 38 individuals in all, is far too small for any definite conclusions to be drawn from it. But the results do point in a particular direction and in the next section I will report on a similar study, with similar results. If the events of infancy prove to be as important to the adult IQ as the above facts indicate, the question of heredity and environment takes on a new aspect. In this case, the limits to intellectual development are set not only by damage to the central nervous system, education and training and prevailing attitudes during childhood, but also by earlier and more subtle factors. It is also true that if there is an upper limit to the intellectual capacity of a normal human being, there is reason to seek that limit in the fundamental attitudes to life and the fundamental conceptual apparatus which are beginning to evolve even in infancy and which form the basis of the mental apparatus which will then be used throughout life.

One of the weightiest facts to indicate that Galton's theory of heredity is wrong comes from intelligence tests on large groups of people in the same community with a longish interval between the tests. If Galton's theory were correct, the average IQ of the population would diminish with the passing of time. Individuals from large families have on average considerably lower IQs than individuals from smaller families. The correlation between the size of family and the IQ is about –0.30. But this applies to the normal population. The correlation is higher for backward people. A Swedish study of individuals with IQs between 50 and 70 showed that the number of children per backward individual was three times as high as the number of children per normal individual. The hypothesis that the brothers and sisters of the backward would have fewer children than the normal population, so that the many children of the backward would be counter-balanced by smaller families among the relatives of the backward, was not confirmed. On the contrary, the relatives of the backward had 2.8 children per mother, as against the 2.5 of the normal population.

Until now, every major population study has shown that the average IQ of a whole population increases with time. Such studies were made in Scotland in 1932 and 1947. In each case about 80,000 11 year-old school children were tested, who had been randomly selected to represent the whole population. Scottish Survey showed an average increase of 3–4 IQ points between 1932 and 1947. There were still greater increases when the results of Army Alpha tests on two million men in the First World War were compared with AGCT tests on twelve million men in the Second World War. A survey of the test results of North American high school children over a twenty-year period showed that their average IQ had risen, although a much larger percentage of the population was attending high school at the end of the twenty-year period than at beginning and one might therefore have expected a general drop in IQ, because more and more pupils came from families with "poorer intelligence genes."

One way of combining these facts with Galton's heredity theory is to point out that the number of children in the family has now decreased. The English psychologist John Nisbet has tried to maintain a belief in a decrease in the real average IQ of the country in this way. But this is a very two-edged defence of the heredity thesis, because we at once ask which genetic factors give the first-born qualitatively better genes than the subsequent children, the second-born better genes than the third, etc. Since no such hereditary mechanisms have been demonstrated, Nisbet's explanation of the facts tends to emphasize rather than diminish the importance of the environment.

The supporters of the heredity theory received powerful support when in 1929 R. C. Tryon published his first results on selective breeding of white rats. "A group of rats was given trials in running a maze; a group of the brightest and a group of the dullest were selected. The bright rats were mated with each other and the dull ones were also mated together. The tendency to rapid or slow learning would thus be selectively cultivated. There would be an 'intelligent' and an 'unintelligent' group of rats, a bimodal division, and therefore two different types. Since the ability to run a maze does not normally play a role in a rat's choice of sexual partner, however, these tendencies are generally more randomly mingled and a normal distribution arises. Most human attributes are distributed according to the theoretical normal curve, which indicates that every attribute is dependent on a large number of pairs of genes."

Through a total of eighteen generations Tryon removed the average rats and allowed the best rats in each generation to mate. He did the same thing with the worst rats in each generation. After the eighteenth generation the worst of the "most intelligent" rats were better at finding their way through the maze than the best of the "unintelligent" rats.

The follow-up to these experiments, using racially pure offspring of the "intelligent" and the "unintelligent" rats showed that what Tryon had succeeded in doing was not in separating genes for low intelligence from genes for high intelligence, but genes which determined a number of more or less distinct qualities. The "intelligent" (maze-bright) rats were "characteristically food-driven (food was used to encourage the rats to run through the maze), economical of distance, low in motivation to escape from water, and timid in response to open spaces." The "unintelligent" rats, on the other hand, were "relatively uninterested in food, average or better in water motivation and timid of mechanical apparatus features." Many different successors of Tryon's rats produced similar results. Tryon had not succeeded in cultivating mental ability but a number of emotional and motivational factors connected with the special maze situation in Tryon's laboratory. Tryon's experiment gave no clue as to which of the two rat strains was more intelligent in the sense of having the best chance of survival in the struggle for existence by the use of its mental powers.

25

FROM TEST RESULTS TO CONCLUSIONS ON HEREDITY AND ENVIRONMENT

From the time of the Second World War to the end of the 1960s it was not regarded as really *comme il faut* to give exact percentages for the relative importance to IQ ratings of heredity and the environment. Most psychologists were content to observe that the mutual connection between the two main factors is so intricate that numerical estimates are necessarily misleading.

From 1969 onwards, when Professor Arthur Jensen of Berkeley put forward his assertion that 70% of the variations in IQ were genetically based, this caution seems to have vanished. Now Eysenck asserts that 80% of the variations in IQ ratings of different individuals depend on differences in heredity. Cyril Burt put the percentage at 77. In 1972 a report was published on several years' studies of intelligence as a function of relationship and socio-economic, cultural and genetic factors ("Inequality – A Reassessment of the Effect of Family and Schooling in America"). These experiments were conducted by Christopher Jencks, Harvard Professor of Sociology, who came to the conclusion that "the genotype explained about 45% of the variance in IQ scores."

It is relevant to note that both Eysenck and Jencks are well aware of the most modern genetic theories and mathematical models for the genetic inheritance of intelligence (Jinks and Fulker, 1970). With identical or very similar equations to work with, they reached values for the genetic factor as wide apart as 80 and 45%. How did this happen?

Well, in order to solve the complicated equations used to reach the above percentages, it is necessary to create, and introduce into the equations, in various ways, numerical values based on theories as to the relative importance and functions of genetic and environmental factors. In other words it is necessary to assume a high proportion of what one is supposed to be proving. Since greater importance is ascribed to the environment in the USA than in England, lower final values are obtained from the American calculations.

If we wish to reach conclusions in the matter of heredity and environment without making any initial assumptions, we shall have to throw all our theories overboard, at least for the time being, and examine the empirical data – the results of tests on human beings.

As we have read, there is a mass of material consisting of test results from groups of individuals with varying degrees of relationship to each other, who were brought up together or in different homes. So we have correlation figures showing the degree of identity of IQ scores for these different groups. Most of these correlation coefficients show that the closer the relationship, the greater the similarity in IQ scores. Jensen and Eysenck are among those who interpret this fact as proof of the great importance of genetic heredity to the variations in IQ. But since closer relationship also entails a greater similarity of environment, all that these correlation data are really saying is that hypotheses which attribute all variations to environment and those which attribute all to heredity may be just as reasonable as a fifty-fifty model.

Here I would like to slip in a word about the doctrine of those who defend the heredity theory. Most geneticists seem to agree that genetic heredity determines our physical structure and appearance, as well as our hormone system and temperament. If they are right, variations in IQ between individuals must of course be to some extent genetically determined. Whether we are slow or quick by nature, whether we are introverts or extraverted, active people, whether we are ugly and repellent or well-made and socially acceptable, whether we are more interested in the opposite sex than in reading books, etc. – all this is to a great extent genetically determined. And all this affects our performance in intelligence tests in various ways.

But when Jensen and Eysenck talk about their respective 70% and 80%, they do not mean this indirect connection between genes and IQ variations. They mean that differences between the genes of the central nervous system result in brains of different capacity and that this brain capacity determines 70% and 80% respectively of the results of IQ tests. The professors mean that the variations in IQ scores directly reflect differences in the constitution of individual brains.

Jensen and Eysenck have remained among Galton's faithful followers. Therefore, in my opinion, they are guilty of a gross simplification of the true circumstances. When the nobility had to hand over its hereditarily privileged positions to the rising bourgeoisie in the 1880s, the vacuum left by the myth of the aristocrat's god-given and nature-given hereditary superiority had to be filled. The bourgeoisie had to be able to justify the privileged position of its own social class. Galton produced the timely idea that a biologically based mental superiority distinguished the élite of this group which had now occupied the leading economic and social positions in the community.

Superiority in relation to the noblest of all human attributes, mental capacity.... The concept of the capacity of the brain reflects the outlook of the 1800s. A brain of a type of which there was perhaps only one in every thousand could retain infinitely more facts, more advanced concepts, more extraordinarily complex networks of ideas than any average brain.

I believe that this concept of capacity is misplaced. The human brain, in the first place, has the power to accept abstractions; i.e. when it has learned an enormous number of individual facts, one over-riding concept is created in the mental system, which comprises in itself the essentials of the individual cases and therefore allows the individual to forget the separate cases and make room for new information without consequently becoming more ignorant, duller. The normal brain has, in other words, a constant, built-in over-capacity, which also functions as a protection against overloading.

In the second place, my understanding is that the human central nervous system can best be compared with a mass-produced computer, i.e. the brains of most new-born individuals function similarly, have about the same number of cells and nerve channels and a more or less equally complex structure. There do not seem to be brains which at the genetic level or in the womb are already significantly better equipped than normal brains for the type of intellectual activity required in intelligence tests. On the other hand, in every hundred or thousand new-born individuals there are a few who have defective brains, less well equipped than the normal brain. It is also quite obvious that a percentage of all adult individuals have brains with defects caused by malnutrition, toxic damage, alcoholism, road accidents, etc.

According to this view, we then have a large number of computers from the same mass-produced series, which are put out to different firms and companies. The companies have different fields of activity (the individual's profession, social group, etc.), different managements which decide how the computer shall be used (the individual himself, influential relatives, idols, etc.), use their apparatus to a different extent (absent-mindedness, day-dreaming). A small proportion of all computers have more or less serious defects and a small number of these defective computers were defective when they left the factory. The variations in performance between the non-defective computers depend almost exclusively on the variations in the programmes with which they are fed and the data in the data banks (the software). Briefly stated, this is the computer model for the relationship between the IQ and genetic and environmental factors.

If this explanatory model for the variations in IQ in a normal population, based on a summary and intuitive assessment of relevant data, is a more correct reflection of the true circumstances than the Galton model, we should be able to find correlation data which do not agree with Galton's model but do agree with this computer model. (Despite the fact that all intelligence tests are carefully composed on the pattern of the Galton model and that therefore most IQ tests give results which reinforce Galton's model.)

The fact is that such data do exist. For instance, there is

the correspondence between the IQs of dizygotic twins who have grown up in the same home. Dizygotic twins are genetically no more alike than ordinary siblings. Yet an impressive number of tests based on a large material show a mean correlation of 0.63 for dizygotic twins, as against 0.52 for ordinary siblings. This higher correspondence between the IQ of dizygotic twins cannot be explained in the context of the customary Galton model. There are two possible explanations: either we can assume that monozygotic twins have been included in the material, owing to the uncertainty of twin diagnosis, and have increased the correlation; but in this case most correlation coefficients from monozygotic twin studies would be too high, since the less similar monozygotic twins are not included, being classed as dizygotic twins. Or else we can undertake the measurement of extremely subtle (from the standpoint of the Galton model) environmental factors and attribute to them a relatively high importance for IQ. The only persistent difference between dizygotic twins and ordinary siblings is that dizygotic twins arrive at the same point in the family life and share the family's social, cultural and economic conditions during the same period of time. Ordinary siblings arrive in the life of the family at intervals of one or more years and therefore occupy different positions in the age hierarchy of the sibling group and do not share the family circumstances under exactly the same conditions.

Since these very subtle environmental factors lead to a 20% higher degree of correspondence between the IQ scores of dizygotic twins, one might expect monozygotic twins to show a considerably higher degree of correspondence between their IQ scores. After all, they enjoy the same special position as dizygotic twins but are also always of the same sex, look alike, have the same temperament and nervous tendencies, have almost identical interests and motives, meet the same reactions and attitudes from those about them, who are often unable to distinguish between them, etc. And as we saw, the correlation coefficient for the IQs of monozygotic twins, when they have grown up together, is about 0.89. (Naturally this does not prove that the computer model is more satisfactory than the Galton model, but neither does it prove the opposite, as many people seem to think.

As with most of the correlation figures in this book, I have given the "raw" coefficient. Usually a "raw" coefficient is converted by mathematical methods which raise the highest coefficients steeply and the lower coefficients considerably less (corrections for unreliability and restriction of range).

So the computer model should be capable of verification by data from tests of highly intelligent individuals. According to the computer model these individuals should have high IQs, not because they have better than normal brains, but because of better programming, more precise memory data which are more relevant (to the test situation), etc. The children of the highly intelligent who have been adopted in infancy should certainly, according to the data model, show a degree of correspondence with the IQ of their parents, owing to the environment of their first months (when the predominant motivation is laid down) and because of selective placement in the adopting families (the correlation coefficient for the education of natural and adopting mothers is about 0.25). These factors together should lead to a somewhat lower correlation between the highly intelligent natural parents and their adopted children than the normal correlation between parents and children in the same home (0.50). But if the Galton model were correct the average IQ of the adopted children should be closer to the average of their natural parents than of their adoptive parents. Unfortunately I have been unable to find studies of this type.

For the converse situation, however, empirical data do exist. According to Galton's model, when less intelligent mothers have allowed their children to be adopted by more intelligent families, the children's IQs should be on average closer to that of their natural mothers than of their adoptive mothers. As we have already seen, study of this type has been carried out which reinforces the computer model and cannot be satisfactorily explained on the basis of Galton's genetic model.

A third powerful argument in the debate as to the most realistic intelligence model is Fraser-Roberts' study of the

IQ of 562 siblings of imbeciles (IQ 30–68), which I referred above. When Fraser-Roberts divided the siblings into two groups (one group of siblings of the most backward, the second of siblings of the imbeciles closest to the normal), he found that the siblings of the most backward had an average IQ which was roughly normal, about 100. The siblings of the second group, that is of the most normal of the imbeciles, had a considerably lower average IQ, about 80. This is, of course, difficult to explain on the basis of the genetic model, according to which the siblings of the imbeciles with the lowest IQ should have a lower average IQ. According to the genetic model, the most backward would have the worst genetic heritage of intelligence, the worst brains. According to the laws of genetics, their siblings should share this weak genetic heritage. (In his book "The Inequality of Man" Eysenck explains this dilemma by departing from his usual model, according to which intelligence is determined by a large number – more than 100 – of different genes with different dominance, penetration, etc. For this special case Eysenck postulates "single recessive genes, or mutant genes, whose effects are so strong as to override completely all other genetic or environmental effects involved in intelligence."

If we proceed from the computer model things are different. Since each individual's intellectual apparatus is consolidated in conjunction with both the close family circle and the outer circle of school, neighbours, friends, all the media and systems of the community – individuals with particularly low IQs must have defective brains. Quite apart from the possible incompetence of their parents as guardians and guides and providers of an adequate childhood environment, the influences from the outer circle would guarantee an IQ higher than that of the parents. In those cases where, despite everything, the IQ is abnormally low, we must be dealing with defective brains. As I have shown, these defects are considerably more often associated with pre-natal damage or damage at birth than with damaged genetic structures passed down from generation to generation. A large number of siblings of such seriously backward people should therefore achieve, on average, the normal IQ scores for the population as a whole.

Some of the less severely backward, those with IQs of about 60, may also have damaged brains. But considered as a whole, this more intelligent group probably has less brain damage than the group of the seriously imbecile. The imbecility must be a result of programming. Since the siblings of these individuals were programmed by the same parents and close circle, while the influence of the outer circle should lead in the direction of a more normal IQ, they should fall on average between the IQ of the least intelligent siblings and that of the normal population, and according to Fraser-Roberts' study that is what they do.

So we have several well-documented and incontrovertible test data which are difficult or impossible to explain on the basis of the genetic or Galton model, but which accord well with the computer model of the relationship of the IQ to genes and environment. However, there are some data which at least at first sight seem to support the Galton model. Psychologists of the Galton school make much, for instance, of the concept of regression, which they consider to be a typically gene-based phenomenon. Regression, as far as the psychology of intelligence is concerned, is the tendency for the IQ of offspring to lie between the average IQ of the parents and 100, the normal figure.

But since an individual's social and cultural environment is at least as much determined by the influence of friends, teachers and neighbours (not to speak of newspapers, films and television) as of his parents, and since this "outside" influence tends to "normalize" the child, to bring his intellectual apparatus closer to the mean value of the population, the observed phenomenon of "regression" is just as well explained by the computer model as by Galton's genetic model.

Then we have the fact that the IQs of adopted children correlate more closely with those of their natural mothers (0.40) than that of their adoptive parents (0.23). However, as I have said, it turns out that an infant and nursing period of some months spent together does transfer behaviour patterns from mother to child which later in life produce slight correspondences in IQ, although the children have grown up in another home after the first few months. Moreover, the child inherits its temperament and appearance from its biological parents and these genetic factors lead indirectly to correspondences in IQ. Temperament and hormone system, appearance and physical structure constitute a good deal of what in everyday language we call our personality, and the nature of the personality decides to a great extent whether a person moves towards intellectual interests or other things, whether he gets on well at school or does better on the playing-field, whether he communicates much with teachers or parents or is more withdrawn – all circumstances which are significant to performance in an intelligence test.

One further empirical research result which at first sight seems to falsify the computer model, but does not in fact do so, comes from Schull and Neel's major study "The Effects of Inbreeding on Japanese Children" (1965). Tests on these children showed a lower IQ, the more closely the children's parents were related. Experience of inbreeding in animals and plants has shown that certain aspects of the genetic material degenerate. The genetic basic material for such more highly coordinated functions as intelligence should therefore be the first to suffer injury. However, the same study showed that the children's performance in different school subjects, including music and gymnastics, had a still stronger negative relationship to the degree of inbreeding and "several types of 'socio-economic' scores, based on parental occupation, education, mats per person in the home and food expenditures per person per month, have been constructed. These all appear to be significantly related to consanguinity, the inbred having the lower score."

A fourth factor which seems to contradict the computer model emerges from a study by R. C. Johnson (1963), which compared monozygotic twins separated at an average age of two months with monozygotic twins who had been together for an average of 24 months before separation. According to the environmental hypothesis, writes Eysenck, reviewing the study, there should be a great correspondence between the IQs of the twins who had been together longest before separation. Instead, the average IQ variation was twice as high (9.4 points) in this group.

In "The Inequality of Man" Eysenck writes apropos of Johnson's study: "It will be remembered that the possibility of some such specially important early period of life together was the main criticism of the 'twins reared apart' proof for the importance of genetic influences; we can now see that this criticism has no force as far as human subjects are concerned." Johnson himself writes at the end of his paper that "the results do not support the idea that similarity in early environmental enrichment or deprivation is related to later IQ similarity in children."

If Johnson's results were correct they would be a two-edged sword, for what genetic factors could explain one group of twins having twice as great IQ variations as the other? However, the twin pairs whom Johnson studied were identical with the 19 twin pairs of Newman, Freeman and Holzinger to which I referred above (except that Johnson added four new twin pairs, each one of which was taken from the reports of four other authors). I showed that the two groups of twin pairs had widely different situations as regards schooling and age which had to be taken into consideration if correct results were to be achieved. When I did this by matching the pairs of twins who had had the most similar school training experience, I divided up these matched pairs into two groups, alike with regard to school and education, but different according to the periods of time the twins had spent together before separation. I found that the twin pairs who had been longest together had the least average differences in IQ. The twin pairs which had spent the shortest time together before separation had almost 40% greater IQ variations between them than the twins who had been separated after a longer period.

The extremely important discovery of the influence of the early environment on learning capacity and (for human beings) the IQ, was first made on the basis of the observation that "home-reared 'pet' rats, from an ordinary strain of laboratory rats, were significantly superior on a learning task when compared with rats from the same strain reared in a normal laboratory environment. A large number of very well controlled experiments have extended Hebb's initial findings."

In contrast to what Eysenck writes and in agreement

with an intuitive commonsense outlook, it is obvious that these early environmental influences are of the highest significance for "human subjects." The fact that Johnson did not succeed in detecting anything but an inverse relationship, that the twins who had been longest together had the greatest differences in IQ, is natural and excusable. Where such small groups of material are concerned, there are a number of special features which often overshadow the associations and factors we are looking for in the group studied.

In this connection it may be interesting to have a look at Shields' big monozygotic twin study "Monozygotic Twins Brought Up Apart and Brought Up Together" (1962). Shields worked with two groups of monozygotic twins. One group consisted of twins who had grown up together in the parental home, the other of twins separated at birth or a few months or years later, who had grown up in different homes. Both groups of twins were tested with two different intelligence tests whose scores were collated in a single intelligence measurement.

Shields' material can be analysed in the same way as I have done above with Newman, Freeman and Holzinger's 19 pairs of twins. Shields' group of monozygotic twins brought up in different homes consisted of 44 pairs. From this material I excluded 8 pairs who had been separated but reunited later in childhood, since it is difficult to assess the effect of this on IQ variations. Three pairs for whom no IQ data were available were also excluded. The remainder, 33 pairs, were divided into two groups. One consisted of twins separated immediately after birth, 12 pairs. The second consisted of 21 pairs of twins who had lived together for at least 3 months (average 2 years) in the same infant environment. The variations in intelligence test scores (here somewhat unrealistically called IQ scores) were converted into percentages in order to make the comparisons more accurate.

According to the environmental hypothesis, the twins who had lived together for the first few months of life should have shown less IQ variations within each pair than the twins separated at birth. As in Newman, Freeman and Holzinger's study of 19 twin pairs, the hypothesis was not confirmed and the twin pairs who had lived together longer had about 15% higher variations in IQ.

A more detailed analysis of Shields' material shows that in 11 of the 33 twin pairs at least one twin was adopted by a completely unrelated family. These 11 pairs were sharply distinct from the remaining 22 pairs, in which the twins, although they had grown up in different homes, were all brought up by close relations from the same social level and often in the same town or village (in general one twin had grown up with his mother and the other with an aunt or grandmother). The 11 "adopted pairs" showed, as we might have expected, a greater average variation, *in fact almost 40% greater than the average variation for the 22 "family pairs."* (This factor is in itself a powerful argument for an environmental model.) Since no less than 8 "adopted pairs" were in the "together group" as against 3 in the group of those separated at birth, the "together group" showed higher average variations for reasons unconnected with the common environment in infancy.

In order to arrive at an accurate figure for the variations in IQ between the two groups separated at different points in time, I therefore had to remove the "adopted pairs." When this was done I had a group of nine twin pairs in which the twins had been separated at birth and a group of 13 twin pairs who had lived together for at least 3 months before separation. The average period together was 32 months. For both groups, the difference between their childhood environment was of the mother's home/aunt's home type.

For the group separated at birth the IQ variations between the twins in each pair proved to be about 20% on average. The corresponding figure for the group of twin pairs who had lived together for the first months of their lives was 16%.

As in the case of Newman, Freeman and Holzinger's 19 twin pairs, I then found that the hypothesis of the importance of the first months of life to the IQ of the adult was confirmed. If further studies of larger groups give similar results, it can be considered as satisfactorily proven that the environment in the first months of infancy has a great influence on the IQ level as an adult. In such studies care must of course be taken that the groups compared resemble one another both as regards home circumstances and schooling, and as regards age (the older the twin pair, the longer the twins have lived in different environments and the greater average IQ variation they will show).

Shields' study reveals many other interesting facts. The separated twins, whether separated early or late, differed markedly from the control group of twins who had grown up in the parental home. The separated twins had IQs about 20% lower on average than the twins who had grown up together. This indicates in the first place that the separated twins were negatively affected by the psychological and physical "transplantation" at the most sensitive period of development. In the second place they must have grown up under less fortunate conditions, at least from the intelligence standpoint. But the spokesman for the genetic model can naturally protest against this interpretation and insist that the parents of the separated group had worse intelligence genes than the parents of the group brought up together.

To summarize, I would like to insist that a conscientious analysis of Shields' material does not justify his assertion that "the importance of heredity for intelligence is confirmed."

Logically and psychologically, the computer model is more satisfactory than the Galton, or genetic model. While there is a good deal of well-documented test data which is very difficult to explain in the context of the genetic model, there are no correlation figures which cannot be explained by the computer model.

As I see it, we should criticize those psychologists who support the genetic doctrine for four cardinal errors inherent in their work in the intelligence field. First of all, they over-frequently simplify the purely psychological causal connections, while ignoring our extremely inadequate knowledge of this subject at present. In the second place, they refuse to take the Rosenthal effect seriously and therefore seldom carry out double-blind studies. In the third place, they make use of statistical, mathematical and mathematical-genetic models which are so advanced in relation to the comparatively coarse test data which they are using as the only directly empirical input data, that they are often dazzled by the precision of the instrument and elaborate the few and coarse empirical data to the point of exaggeration; in other words, they obtain results which externally appear extremely accurate and reliable, but which in reality say little more than raw correlations and test scores. In the fourth place, they generally refuse, on the basis of their desire to defend the Galtonian legacy of ideas, seriously to try to clarify and understand the intricate network of subtle connections which many psychologists believe they can perceive between the test result and the personality and various aspects of the circumstances of growing up. While the spokesmen of the heredity doctrine cheerfully accept the idea that "more than 100" different genes in a complicated network of relationships together determine the IQ, none of them, as far as I know, recognizes the possibility of "more than 100" distinguishable environmental factors which together, in a complicated network of relationships, programme the growing child.

Here are some of the circumstances and attributes which have been found to vary to a greater or lesser (but always significant) extent in relation with IQ. The order in which these items are listed does not reflect any classification of the inherent significance of the various factors.

Height of the individual
Father's profession
Number of years in school
Number of siblings (negative relationship)
Position in family
Social group of parental home
Father's economic status
Average TV viewing (negative relationship)
Average book-reading
Self-confidence according to attitude scale measurement

Age (negative relationship, applies only in adulthood, except for country children)
Degree of authority in parental home (negative relationship)
Criminality (negative relationship)
Alcoholism (negative relationship)
Mental disease (negative relationship)
Emotional adaptation according to tests
Degree of parental rigidity (negative relationship)
Parental ambition
Mother's education.

Perhaps the most significant environmental factors are the series of events which lead to a child beginning to take an interest in certain things rather than others, the reinforcement mechanisms and deterrent factors which lead the child along a path which will reach its optimal point when he has become a mature, occupationally and socially established person; also the series of successes and failures in the attempt to control his situation in the world which leads the child towards increasing self-confidence or to a resigned conviction of his own incompetence.

26

INTELLIGENCE AND CHILDHOOD ENVIRONMENT

Obviously, we want to have intelligent children. Even if an intelligence test measures only a little of an individual's knowledge and ability, no one will have the freedom and opportunity to reach any position he wishes in our community if he scores less than about 120 in any IQ tests he may be given. A lower IQ score in any test means that he is not in command of our society's most important symbol system and that he will therefore be unable to obtain good enough school reports, to communicate in a sufficiently "intelligent" way, and have sufficient intellectual self-confidence to confront all the intellectual problems he may meet in school and working life.

In the private sector in the society of the future, the trend will probably be towards an attitude which makes formal intelligence less important. Rather, the individual's power to influence events and other individuals will be the essential factor. School reports, IQ scores and school leaving certificates will fade into the background by comparison with such questions as "can this man sell more of my refrigerators than that one?" "can this woman compose advertisements which will really make people buy Bluff-Bluff?" One reason why we can expect things to develop in this direction is that many people in the community will have a high formal intelligence because of the development of the school system and the standardization of the language through television.

This trend is counteracted, or rather, counterbalanced by the fact that society's state and communal institutions are going to expand and absorb more and more people. Much of the work in institutions follows formal rules. New analyses and instructions have to be produced. Men and materials must be directed according to a prior plan. In this sector of the community the demand for formally intelligent individuals will increase.

The factors in the childhood environment which enable the child to gain high scores in intelligence tests are often the same as the factors which make him a creative and independent individual. In this chapter I will be reporting on various experiments and studies which indicate some of these factors.

Many experiments have been carried out on so-called sensory deprivation. This means that the normal sources of sensory impressions are removed, for instance by isolating the individual in a room without light or sound, or by lowering him in a vessel into a liquid of the same viscosity as the body tissues, in a dark, sound-proof room. The only sensory impressions which a person in this situation can receive are those from his own internal organs.

Dr Jack A. Vernon has carried out experiments in which his experimental subjects were shut up separately in dark, sound-proof rooms. When sensory deprivation of this kind lasted longer than a few hours (the length of the periods of isolation was between 24 and 72 hours), the experimental subjects began to display a greatly diminished ability to solve complicated intellectual problems. On the other hand, their ability to learn things by heart increased.

The experimental subjects also became abnormally sensitive to suggestion. Their ability to think was clearly affected. Their power of concentration weakened. They also had great difficulty in dropping any method of problem-solving which proved ineffective. In other words, they became rigid and tended to persevere with defective ideas once they had received them. (Another interesting result was that most of them lost weight appreciably, although Vernon thought the opposite would happen.)

Dr Vernon's experimental subjects regained their normal mental capacity shortly after being released from their isolation cells. However, there is reason to suspect that long-term sensory deprivation for a child in infancy can result in diminished mental capacity for longer periods.

W. Goldfarb, the psychologist, compared a group of young people who had spent their first three years in a children's home, that is in a very sterile and uniform environment with few personal relationships with adults, with a group of young people who resembled the group from the children's home in many essential respects such as age, sex, etc., but who had grown up in adoptive homes from the first year of their lives. The IQ score of the group which had spent the first three years in children's homes was 20 points lower, they had poorer memory, less ability to form concepts, less planning ability and were backward in speech development. They were also more anxious and more aggressive than the ones who had grown up in families. They were over-active and found difficulty in concentrating. Many of them also showed signs of what is often called psychopathy – an emotional coldness and indifference.

It is interesting to note that other studies have shown that children from families in which they were treated in an inhuman way increased their mental powers after being placed in children's homes. Of course, it is not the children's home as such that causes mental retardation; it all depends on the attitudes of the staff and the environment in general. Unfortunately it is quite usual for a children's home to involve an environment of the type described in the book "Miljöterapi" ("Environmental Therapy") by three Swedish doctors, three psychologists and a sociologist:

> "The 'children's department,' on the hospital pattern, is often in a long, pavilion-like part of the building with a corridor down the middle. Administrative offices on one side of the corrider and on the other a row of children's rooms with 3–4 barred beds in each. The corridor wall is generally made entirely of glass, so that the children can be constantly observed from the outside. The 'staff department' with despatch, restaurant, day-room, lecture-room, etc., is generally situated in quite a different part of the building. Neither the children's cries nor their shouts of joy penetrate there, the people there are at a great distance, both spatially and psychologically, from the nursery environment. The picture is children there, grown-ups here.

> "If we go back to the children's department and open the glass door to the children's room we are once again surrounded by the same anonymous atmosphere that we met at the entrance. The beds are generally in the corners of the room, all of the same type and colour; the bed-clothes are uniform. The wall decoration is generally conventional and standardized, the colours are soft-toned but also often drab. The toys are pedagogically correct and suitable for the various age groups, but they are generally few in number, unimaginative and lacking in variety. The toys are stored away at rest time and are never lying about the place, either in the department or outside the house. The dominant impression of the children's room is that in any case it lacks all resemblance to a real nursery. On the other hand, it bears a strong resemblance to a hospital ward. One has the feeling that order and cleanliness, hygiene and neatness are the principles on which the children's living-rooms are designed."

In 1949 René Spitz published the results of an interesting study on how infants should be treated in order to have a well-developed intellect. Spitz investigated infants in a nursery and a foundling home. The two institutions resembled each other in that children came to them new-born, the food was nourishing, the premises and hygienic conditions satisfactory. The important difference was that the children in the nursery were cared for by their own mothers, who came there every day, while children in the foundling home were looked after by paid staff, with up to 12 infants each in their charge. A development quotient, a DQ, was worked out for each child on the basis of various tests. DQ corresponds in a way to IQ, but is calculated, owing to the youth of the child, along different lines from the IQ.

Figure 44 describes the DQ of the children in the first years of life. The DQ for children cared for by their own mothers in the nursery remained at about the same level throughout the first year of the child's life. The DQ for the less personally handled children in the foundling home was

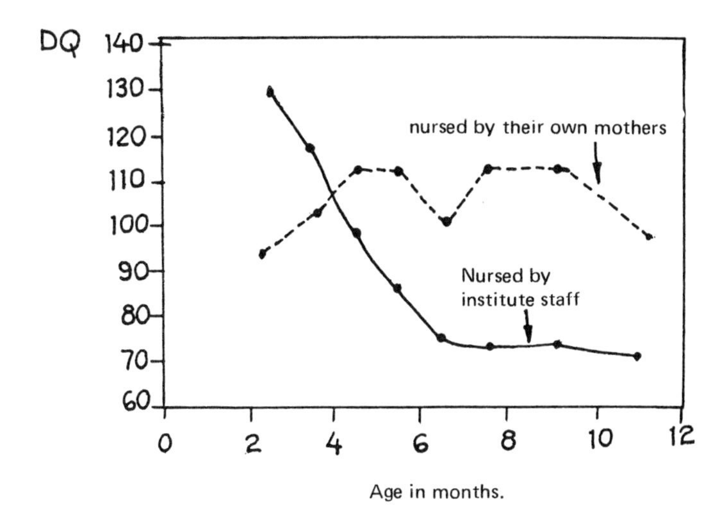

Figure 44. Development of the infant in the Spitz study of the first twelve months. Children in the Nursery compared with children in the Foundlinghome.

higher than the rest to begin with, but dropped until after 6–12 months they were about 40 DQ points below the others.

Another interesting observation when the children were followed up over a five-year period was that none of the children in the nursery died. More than one third of the children in the foundling home died, most of them during their first year. Now, it is quite possible that the children in the foundling home were less healthy to begin with. But this can hardly explain the whole of the wide difference between the mortality rates of the two groups of children.

There were differences in the behaviour of the mothers at the Nursery towards their children. Many mothers left their children for long periods and the children were than cared for by the staff at the home. Spitz studied the effects of this on the children's IQ. Firstly, he found that children who had been separated from their mothers towards the end of their first year lost awareness and interest in their surroundings. During the period of separation from the mother the DQ dropped steeply.

Secondly, Spitz found that if the mother returned to her child within five months the DQ rose so steeply after the mother's return that the earlier fall was compensated. If the separation lasted longer than five months, however, there was no difference when the child's contact with the mother was resumed. The child was irrevocably retarded and remained so throughout the five-year period during which Spitz followed their development.

There is no way of knowing the significance of these facts about infant development to their IQ as adults. Probably the child's general curiosity and appetite for life, and therefore indirectly its intelligence, can suffer permanent damage through an early, long-term separation from its mother. Naturally, it makes no difference whether or not the separation is from the child's natural mother, or, at least, so the data indicate. It is the separation from the person whom *the child himself experiences as his closest companion and protector* which is essential.

There is good reason to claim that a varied and emotionally safe early childhood with a great deal of skin contact with other people is a necessary condition both for purely formal intelligence and for mental capacity as a whole.

Formal intelligence should not be the prime goal for people who want intelligent children: the result may easily be a neurotic attitude to intellectual activity in the child. The child must be allowed to play its way into insights and confidence in its own ability to work out the consequences of various modes of action. In 1956 Wittenborn published a study of 195 adopted children and their adoptive mothers. The mothers were interviewed and placed, on the basis of the interviews, on scales of different qualities significant to their relationship with their children. It was found that the level of the child's intelligence did not correlate very closely with the mother's level of educational ambition for her children. Wittenborn found stronger correlations between neurotic tendencies in the child, e.g. exaggerated aggression, compulsive behaviour and phobias, and the mother's tendency to push the child away and punish it.

The atmosphere in the child's home, the parental

Table 3. Changes in a child's IQ over a three-year period in relation to the psychological home environment. (From an American study.)

Home atmosphere in relation to child	Changes in IQ over a three-year period
Actively hostile	−1
Passively uninterested	−0.5
Dominant-possessive	+0.5
Allowing freedom but little emotional contact	+5
Allowing freedom with emotional contact	+8

attitude to children and to the community and life as a whole have been investigated in various different studies. The data in Table 3 are representative of the results.

There is not much to be said about the homes where the children are not welcome, where they are regarded as a burden and a necessary evil. Homes like that are not very common. On the other hand, there are homes with a dominantly possessive atmosphere. Such home environments are warm, but often so protected that they impede the child's discovery and investigation of his surroundings. The parents are frightened that the child is going to hurt himself and have difficulty in tolerating the child's own fumbling attempts to do things for himself. This type of home regards the social conventions which prevail in the community as extremely important and the child is therefore pressured into performing better than other children as regards such things as learning to walk early, talk, go to the lavatory on his own, read, write, etc. The overall effect of this type of home environment is not much better for the IQ than the effect of a more hostile or passively uninterested environment.

The freedom to investigate and manipulate his surroundings is the most essential factor in the development of a child's thinking capacity. In order for this freedom to be exploited by the child, he must know that he has a safe refuge if the results of his explorations become uncomfortable or downright catastrophic. If this almost purely emotional security is lacking, the child will not dare to use his freedom to explore and manipulate his surroundings.

There is a strong negative connection between the IQ and the degree of authoritarianism in the home. Parents who try to mould their children to fixed patterns, who punish them physically (by physical punishment I mean deliberate, considered use of force, not the odd slap delivered out of pure temper), who will be obeyed at all costs and who themselves like to have a superior authority to obey, are on average less intelligent than other parents. The IQ of children of these rigid, authoritarian parents has a tendency to decrease as they grow older. The opposite is true of parents and children in a more democratic environment.

Formal intelligence should be a by-product of the child's relationship to his surroundings. One day little Peter suddenly shows a lively interest in counting and puts a lot of questions to his parents. If his parents ignore little Peter's questions he will generally lose interest in that department of life quite quickly. But parents can get tired of answering questions, especially if they themselves are not particularly well informed in that field. Then it is a good idea to buy an easy reading book and give it to little Peter (if he is old enough to read). But then if little Peter's parents, in their enthusiasm for his growing intellectual interests, begin to *demand* that little Peter read his nice new book, which in any case cost them quite a bit, there is a great risk that little Peter will soon lose interest in the book and also in counting in general. It may even happen that quite accidentally the book *gets left behind* somewhere or *gets torn up*. One should never, so to speak, *take over* a growing interest in one's children and make it one's own. This is one of the best ways of smothering a child's own spontaneous, independent interest.

All in all, it is true to say that it is the child's own spontaneous interest which must be supported with attention and admiration when this interest leads to new insights for the child. "But if my little Kenneth is never interested in learning to read for instance . . . if he gets left behind at primary school . . . ." Yes, in that case one must notice what he really is interested in. Maybe he is a little interested in playing with dolls. Perhaps he collects dolls and furnishes little dolls' houses. If we notice his own real interests and help him to make contact with others who share them then for instance correspondence with someone in another town may be the gateway to reading and writing ability. At first his mother will have to help with the writing, but gradually Kenneth will find that it is in his own interest to be able to read and write, and from the moment when little Kenneth sees this and asks his first questions about letters and written words, only a little patience and kindness will be needed for Kenneth to write his own name, the dolls' names, the pen-friend's name and in the end a whole letter.

And never say, "Penny doesn't understand that. She's not inclined that way. Charlie, on the other hand, is a very good reader." To implant such ideas in any of one's children is equivalent to laying the foundations for an intellectual inferiority complex. It is unfortunately only too easy to give a child an attitude to independent thinking and intellectual activities which makes him give up even *trying* to understand, makes him reject confrontations with intellectual problems and devote himself to activities in which he feels more competent.

And so to independent thinking. By doing some violence to reality one can divide parents into two opposite categories. The parents in one category solve any problem a child may encounter in advance. Before little Andrew has even managed to say, "Mummy, how am I going to get to school, I'm a quarter of an hour late," his mother has said, "Well, the car isn't at home and your bike needs mending. I'll ring up Aunt Elizabeth and ask if you can borrow John's bike and then you'll be there on time."

A mother in the other category might say: "Well, so you're late. What do you think you could do about it?" And if she has had the same attitude throughout little Agnes' lifetime, little Agnes will work out all by herself that although the bus has gone and the car is not at home and the bike needs mending, Uncle Leslie next door has a bike in the garage which he is not using. "If I run to the telephone and ring Uncle Leslie he'll lend me his bike and if I'm quick I'll get to school on time."

It is obvious which category of parents will have the more intelligent children. This does not mean that parents should not communicate with their children about anything and everything, quite the contrary. But the communication should consist of "Now, *what do you think*, Agnes?" rather than "You must do this and that and the other, Andrew. Remember, Mother always knows best."

A study at the Fels Institute, Ohio, is relevant in this connection. 140 children were tested at yearly intervals from the ages of 3 to 10. As expected, the IQ of some children rose, while others fell. The 25% of the children whose IQ had risen most and the 25% whose IQ had fallen most were selected for closer study. Their personalities were studied by means of tests and interviews.

The children whose IQ had risen most had more initiative, were more independent and active than the lowest group.

If you want intelligent children you must let them use their intellect independently from the very beginning. What does it matter if the child misjudges situations and causes slight inconvenience and perhaps some unintentional material disturbance, or if neighbours and teachers complain that one's child behaves as if no one were really looking after it? An intelligent, independent and creative child is worth a great deal. But it is also important not, so to speak, to *shove* the child into difficult situations in order to "train" it into independence and thinking for itself. Unless the initiative comes from the child himself, there is a great risk of causing a neurosis.

In 1951 Deiter published a study in which he gave the same task to three groups of children. They had to translate long passages of text from code to ordinary writing. They had a code key to help them. There were two ways of solving the problem: either you could translate letter for letter, looking up each symbol in the key, or else you could work out the system of the code, which was very simple. Each code letter corresponded to the letter three places before it in the alphabet. The work went much faster if you discovered the system.

One of the groups was told that it didn't really matter how they solved the problem. The other was told that the problem was important and they would get a nice reward when they had completed it. The third group was told that the problem was extremely important and that they would get a very special reward for their specially important work.

So the three groups started with quite different motivations. It was found that the group with the lowest motivation and the group with the highest motivation worked on the problem without grasping the system. The most highly motivated group kept stubbornly at the task and toiled through the de-coding letter for letter at the highest speed they could manage. Most of the unstressed group with the intermediate degree of motivation discovered the system and finished the task long before the others.

Similar results were obtained in animal experiments. If the motivation is very strong, for a dog, for instance, his ability to think is clearly reduced. If you take a hungry dog and put a big lump of meat in front of his nose on the other side of a high fence, the dog does not realize that it could run a little way along and round the end of the fence and get to the meat that way. Instead, it tries to dig its way under the fence, even if this is completely hopeless.

Experiments like these lead to the conclusion that a child's intelligence develops best if it is not under pressure to achieve. If it becomes too important for a child to do well in arithmetic at school, there is the risk of its sitting patiently and earnestly toiling through figure after figure in a completely mechanical way, without really grasping what it's all about. This will ultimately lead to the child's being unable to deal with mathematics at the more advanced stages, where mechanical manipulation of symbols is unimportant and appreciation of the mutual relationship of the symbols in the system is essential.

The child has to spend a good deal of its daytime life in school for over ten years. What happens there, or more accurately, *how the child experiences* school is basically of greater importance to the child's IQ than any other factor. If the child starts off in the better half of the class, or in the large group of average pupils, no special measures may be necessary. But if the opposite happens – if the teacher decides that unfortunatley little Peter is rather behind his age group – then owing to the structure of the school system itself there is a risk of his getting left behind as relatively less intelligent than his contemporaries to the end of his school career. In school the child who is intellectually superior from the start is given more attention, more distinction, more respect than the others. Even though the trend has recently been moving in the right direction, including the abandonment of marks in primary school, that is how things stand, and will continue to stand as long as thirty different individuals are forced to apply themselves to the same subject regardless of their own interests.

The child who lands from the start in the section of the class regarded as the least intelligent often experiences his schooldays as a torment and may solve the problem by playing truant. A study made in Malmö by Dr. Johan Öbrink agrees with other studies in showing that the most notorious truants have lower intelligence test results than other pupils.

If a child has difficulties in school and is in that part of the class which receives the lowest marks, he will obviously not be interested in school and school work. The reason may basically be of two kinds: either he is not interested in the intellectual activities which go on in his class, or he may have difficulty in understanding what the teacher is getting at. If the first reason applies it will be difficult to do anything about it in our society. The school system is, after all, compulsory, so you have to be inventive, or have power and influence.

But if the difficulties result from the child's not understanding, then the first thing to do is to make the child *understand* that it will be able to understand. So we have to look for activities in which the child in one way or another realizes that he can do as well as other children. And then we have to go back through the symbol system, whether it be letters or figures, to the point just before the first "I don't understand that." For instance, mathematics is a game with certain symbols: figures, mathematical signs, brackets and so on. There is nothing mysterious about them. If you know the exact meaning of each symbol and the rules which apply to it, you will understand mathematics. But if you have missed the exact significance of any of the basic symbols or rules, you are never going to understand, and maths will remain a mystery even if you learn mechanically to deal with a few standard calculations.

It is quite obvious that it is better for a 17 year-old to leave school without knowing the multiplication table, but knowing that if he cares to learn the multiplication table he can do it in a few hours, than for him to leave knowing his tables but believing that he has less mental capacity than his contemporaries.

In the section on intelligence and position in the family there are reports of many connections between criminality and upbringing. Studies of more than 10,000 juvenile delinquents in the USA have shown that these offenders have a considerably lower IQ than other juveniles. A study of 1,800 individuals with criminal tendencies produced an average IQ of 85 (Shakow and Millard). This means that most methods of upbringing and parental attitudes which are correlated with felonies by the young also lead to slower intellectual development.

New environmental factors such as television also have a relationship to IQ ratings. In a study published by Lotte Bailyn in 1959 she shows that the children who spend most time in front of the television set have a lower IQ, on average, than other children. But watching television is also related to the parents' social group. Children from higher social groups spend less time, on average, in front of the television than other children. So it is not possible simply to say that TV seems to have an adverse affect on intelligence.

Finally I would like to quote the conclusion of a report on a study of highly gifted children carried out by the psychologists Mellinger and Haggard in the late 1950s:

> "The results of this study therefore support the theory that 'learning capacity', 'memory', 'intelligence' and 'personality' are in no way isolated and independent processes but rather represent various expressions of an individual's attempt to deal with and adapt to reality *as he experiences it*."

27

INTELLIGENCE AND CREATIVITY

In this book I have used "intelligence" to signify what is measured by intelligence tests. Intelligence in this sense is a relatively unimportant attribute in comparison to creativity. Creativity can be defined as the capacity to make the events one desires actually takes place – the creation of actions or things.

With this definition of creativity, the intelligence test can be said to measure creativity to the extent that the testee can be assumed to want a high score in the test. But the desire to get a high score in the test is a difficult thing to pinpoint. Many testees, especially from lower socio-economic groups in the community, do not care how they get on in the test. In general, test psychologists have assumed that the desire to get a high score is largely the same for all testees.

But the ability to ensure that he achieves a high score in an intelligence test is only a little part of a man's creativity. Creativity applies to all spheres of life, from the capacity to make a good stew to the ability to work out a contract. As a measure of creativity in the widest sense, the intelligence test is more or less worthless. It represents such a small part of everything that goes on in the world. What inhibits a person's creative capacity is in the first place *his own idea of what he can do and his idea of what is acceptable to others*. A person may have ideas about his intellectual inferiority, or have insufficient knowledge, be too slow-thinking to achieve high scores in intelligence tests. And yet he may have an extraordinary power to create and realize ideas and to think in contexts where there are no time limits. On the other hand, many people of high formal intelligence lack the capacity to create. They have an excellent ability to read and understand instructions and directions, they are conscientious and useful in this way to the community (assuming that the instructions and directions the person obeys are concerned with something which in the long term increases the sum of human happiness).

It must be obvious that creativity cannot be measured by tests, since any creativity test can represent only a small proportion of all the possible areas of activity in the world in which creativity can be shown. So every creativity test must consist of tasks given to the testee, and in order for the result of such tests to express something about an individual's degree of creativity compared with others, a large number of individuals must be given the same tasks and their reactions to the test must be assessed according to the same norms. As soon as such common norms are established the test loses much of its creative aspect. By then it has already been decided in advance what type of results are a sign of the creative gift. There will be no question of creating something new, but rather of adapting to existing norms.

28

ARE INTELLIGENCE TESTS NECESSARY?

It may look as if the nature of the intelligence test and the uncertainty of its results and the intimate relationship between the results and schooling and socio-economic environment make them superfluous. This is not the case.

If we drop the idea that intelligence test results are predominantly genetically based, both the test and the concept of intelligence are extremely useful. There is no better method than the group intelligence test for a quick assessment of a person's ability to understand and obey written instructions and to manipulate the common symbols of our culture.

Moreover, when we use inter-cultural tests which are knowledge and value-free, such as Raven's Matrices, we can make quite objective comparisons between individuals' capacity to think (think in terms of geometrical relationships). For posts in institutions where the individual's creative capacity is unimportant compared with his ability to carry out given orders, the intelligence test is a practical selection tool.

29

THE CONCEPT OF INTELLIGENCE

The concept of "intelligence" is useful only when it is freed from the burden of Galton's genetic postulate. It can be used in the same way as before, that is, to define a person's ability to think in terms of concepts and values which are fundamental to Western culture.

But if we prefer to use "intelligence" so that the word has the same, or almost the same meaning as the ability to think in general, we must take the trouble to change most of the texts where "intelligence" occurs (at least the scool textbooks) and begin to use the word in quite a different way from before. Since this is a nuisance and would mean a major change in our speech habits, it is more convenient to go on using "intelligence" as before, but when we do so we must be quite clear that in this sense Westerners *are* more "intelligent" than people from other civilizations.

What we must beware of is confusing intelligence with the ability to think in the widest sense, regardless of the culturally based conceptual structures and value norms of such thinking. Until a definite linguistic habit has been formed it will be advisable to define the word "intelligence" in express terms when we use it.

APPENDIX

Validity criteria

Up to 1970 the Stanford-Binet test enjoyed a position as a national standard of intelligence corresponding to that of the prototype metre in Paris as the standard measurement of length. Every new intelligence test was validated by comparing its results with those of the Stanford-Binet test.

But intelligence researchers have become aware that it is unsatisfactory to have no validity criteria other than the existing test and statistical postulates. This has led to the emergence of a host of different concepts of validity. The commonest are:

Face validity, i.e. the problem *appears* to require intelligence for its solution.

Content validity, i.e. how well does the problem correspond to some crtierion of what such a problem should contain. In practice this means that in essential points the problem must resemble problems in previous tests.

Predictive validity, which is the most important validity concept because it is not as up in the air as the others. If an intelligence test result shows that people with high scores later achieve high marks and academic distinctions at school and university, the test is said to have high predictive validity.

Concurrent validity, which means that teachers, officers and others are asked to assess the individual's intelligence in relation to other individuals in the same group. The assessors' ranking of the individuals is then compared with the ranking given by the intelligence test result. If the assessors' ranking agrees with the ranking based on the test, the test is said to have concurrent validity. The correlation between the assessors' ranking and that of the test is usually about

0.50, but may be higher if groups including both feeble-minded and highly educated people are used. Another form of concurrent validity means that you take a group of people you know to be unintelligent and a group you know to be more intelligent and give them a test. If the test can separate the intelligent from the unintelligent just as well as the people who were responsible for choosing the individuals in the groups, the test is said to have high concurrent validity. Ultimately, of course, it is people who assess and classify one another.

Construct validity. The term "construct" refers to the fact that intelligence is a theoretical construction and that a good intelligence test must give results corresponding to what one would expect with regard to the theoretical construction, the *construct*. Some of the things one expects have already been mentioned: the results of the test must be normally distributed, they must not display any sexual differentiation and they must have a certain constancy.

Factorial validity. Factor analysis is based on the assumption that intelligence consists of a number of different factors: a memory factor, a speech factor, a numerical factor, etc. A problem in a test is said to have high factorial validity if it has a high correlation with other tests or problems which are considered to measure that particular factor. If we have a test which is considered to measure the numerical factor, e.g. a test which consists exclusively of various arithmetical problems, and if we are trying out a different sort of arithmetical problem from the ones in the test, the new problem will have high factorial validity if the people who handle the new problem best are also best at doing the test, and if those who perform worst with the new problem are also worst at the test.

Validation of Army Alpha and WAIS

The Army Alpha test, which was administered to almost two million men during the First World War, was validated in this way: the test was given to over 5,000 soldiers and several hundred students, schoolchildren and mental defectives from various institutions. Then data was collected as to school reports, teachers' assessments of inherent talents, officers' assessment of soldiers' ability, the rank of the soldiers (!), their skills as evinced during military training, how long they had been to school as civilians, etc. With regard to the mental defectives the criterion was that they should quite simply have the lowest score. After this the whole group was given the Stanford-Binet test and all the correlations were worked out. They were found to lie between 0.50 and 0.80.

In 1939 the psychologist David Wechsler introduced a new intelligence test for adults. It was revised and expanded in 1955 and has since been called the Wechsler Adult Intelligence Scale, WAIS. A similar test series was also designed for children, the Wechsler Intelligence Scale for Children, WISC. It might be interesting to see how an American standard work on tests assesses the validity of WAIS.

First it says how well the results of the sub-tests agree with the whole test, and that the correlations indicate that WAIS, taken as a whole, has a satisfactory degree of construct validity.

Then the correlation with the number of years in school is given. This is about 0.70 and is regarded as satisfactory. Those who had been to school longest thus had higher scores than those who had been for fewer years.

Then the change in the score with increasing age of testees was reported. It was found that the average score for a large representative group reached the maximum just before the age of 30 and then fell slowly. These results were in agreement with existing psychological theory, except for the fact that the maximum in the 25–29 year group was higher than the one reached in testing with other tests. (The age of maximum intelligence for large groups was 16 according to Stanford-Binet 16; about 18 according to Stanford-Binet 37.) The next criterion of validity is the pattern of IQ distribution. The two characteristics of a valid psychological test are that they must give an adequately wide range of variation, so that they cover the great variations in human capacity, and that they must give a distribution which is close to the normal curve. WAIS is described as satisfactory as regards these criteria.

Then follows the correlation with Stanford-Binet and other tests. When validating a new scale for intelligence measurement it is common practice to correlate the results obtained with the new instrument with the results of Stanford-Binet for the same individuals. The reported correlation coefficients lay between 0.60 and 0.90. The correlations were satisfactory, but it was unfortunate that most studies which had compared these two instruments had used non-typical groups, such as hospital patients, clinical cases and prison inmates, by-passing individuals who fell into more normal categories.

Then comes predictive validity. WAIS is compared with teachers' reports, resulting in an average correlation of just over 0.50, then with groups of known intellectual status, consisting of mental defectives, border-line defectives and normal individuals. It was necessary to confirm that each sub-test was distinct, as regards frequency of solution, in the different groups, and this was done. Finally there was a report on a study in which it was found that the correlation between WAIS and measurements of scholastic success for a group of students was higher than the correlation between a group test and scholastic success for the same individuals. The correlation between WAIS and the criteria of scholastic success was 0.53.

The last criterion of validity is *Item Difficulty*. This means that the most difficult tasks in the test cannot be solved by more than a few individuals, while the easiest can be solved by almost everyone. On this point there were some criticisms of parts of WAIS, since with such a high percentage managing the most difficult problems, if this study group were typical, these problems could not differentiate sufficiently well between individuals on the lower levels of the distribution.

Intelligence and the normal curve

The normal curve is the name given to a random distribution curve. If you throw ten coins into the air

10,000 times and write down how many times only one coin turns up heads, how many times two turn up heads, etc., you will observe the following: in 7% of all throws, only one or two coins will fall with the head uppermost, in 24% of all throws three or four will fall head uppermost, in 38% five or six will be heads, in 24% seven or eight and in 7% of all throws nine or ten coins will fall heads up.

Most psychology textbooks maintain that intelligence is distributed accrding to the normal curve. This is true if we mean that the results of intelligence tests are normally distributed. And this is because the degree of difficulty of questions in the test has been so adjusted that the frequency of answers from an average population follows the normal curve.

An intelligence test was constructed in this way in order to agree with Francis Galton's ideas about the distribution of human beings' inborn capacity. Galton did, of course, know that body height in a normal population of adults of the same age was distributed according to the normal curve. Figure 45 represents the body height of 8,500 men.

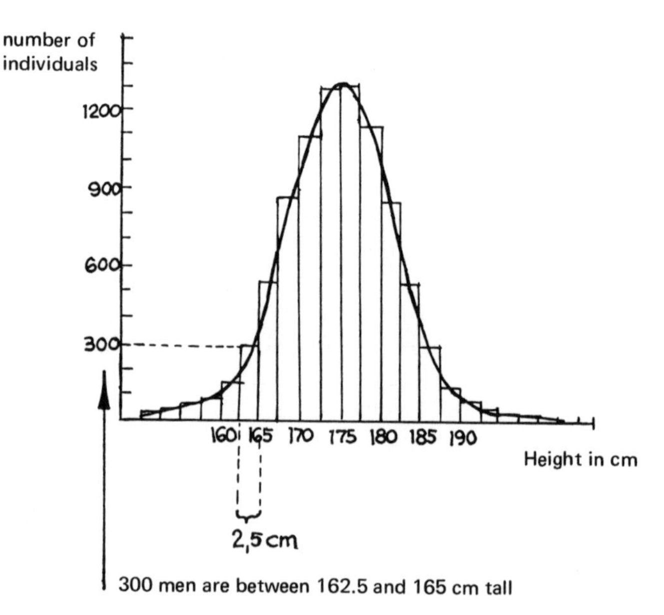

Figure 45. Body height of 8,500 men aged 25.

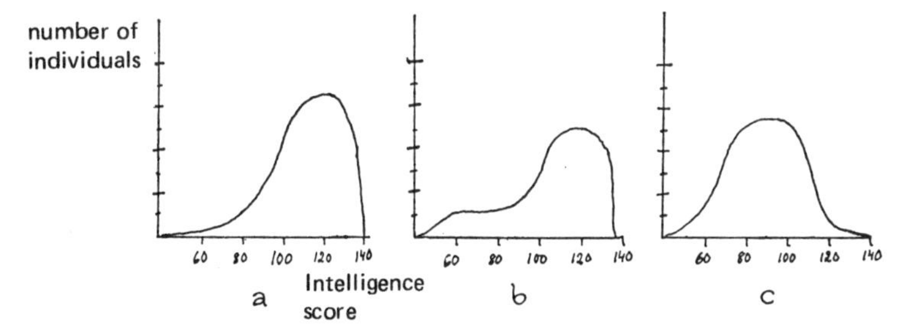

Figure 46. Distribution of intelligence in a normal population when the degree of difficulty of the test problems is altered in three different ways.

The US psychologist Professor Tyler writes: "It is quite possible to change a distorted distribution to a normal curve simply by making the test on which it is based a trifle more difficult or a little easier . . ."

By altering the questions a little – or the time limits in timed tests – one can therefore make the results of intelligence tests follow many different types of distribution. Figure 46 shows three examples of possible different distributions of intelligence in a normal population.

If we investigate various psychological and physiological qualities carefully we find that many of them are distributed quite differently from the normal curve. One

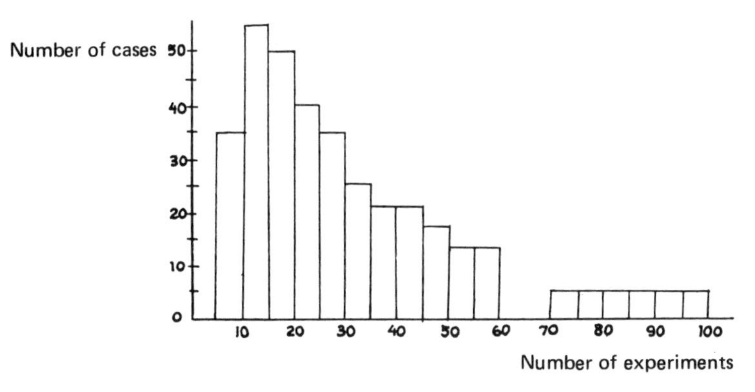

Figure 47. The variation in the ease with which the blinking reflex can be caused in different people. This is an example of an attribute which is not distributed along the normal curve.

example is the distribution of the ease with which people can be taught to blink atuomatically when they receive certain sensory impressions. This is shown in Figure 47.

How the Stanford-Binet test was constructed, and the IQ concept

Binet's 1911 test is the basis of the subsequent development of intelligence tests. There were five different major translations and revisions of Binet's test up to the 1920s but the most important was the one performed by Professor Lewis Terman at Stanford University. He took the 54 problems from Binet's 1911 test, 5 from earlier Binet tests, 4 from American tests and himself composed 27 new problems.

His most important criteria of validity were purely statistical. A problem for 7 year-olds, for instance, should be answered correctly by about 77% of all 7 year-olds. 50% of all 6 year-olds and 95 of all 8 year-olds should also be able to answer the question correctly in order to be included.

Here is an example of problems from Stanford-Binet 1916. They are intended for 9 year-olds:

1. Give the date: day of the week, month, number of day in the month, year.
2. Give the difference between these weights: 3, 6, 9, 12, 15 ounces.
3. Give correct change for small sums.
4. Repeat four figures backwards.
5. Make a sentence including three given words.
6. Give rhymes for three given words.

Alternative 1. Name the months of the year.
Alternative 2. Add together a number of 1 cent and 2 cents stamps.

(The last two problems were alternatives which the examiner could give if he had reason to think that any of the usual problems were unsuitable.)

In connection with the 1916 revision of the Binet scale,

the concept of the IQ was introduced in earnest. In his 1908 test Binet had assembled the problems in eight groups based on the age at which the tasks were performed by 75% of the children. If a child could do all the problems, or all but one, for 7 year-olds but did not perform all the tasks, or all but one, for 8 year-olds, he was was said to have a mental age of 7.

The intelligence quotient is now defined as the quotient of the mental age and the real age. So the convention was introduced that the quotient should be multiplied by 100, in order to avoid the inconvenience of decimal points and still have a sufficiently exact expression. A 6 year-old who succeeded in passing all the tests, up to and including those for 9 year-olds, had an IQ of 9/6 x 100 = 150. It was found that the distribution of IQs in a representative group agreed usefully with the normal curve, although it was rather crowded in the middle. The standard deviation was 12, as against 16 for a more normal distribution.

Where adults were concerned the IQ could not be calculated by means of their real age. The 1916 Stanford-Binet ended with the test for 14 year-olds, plus a test for normal adults and a test for superior adults. Then Terman introduced the convention that adults should reckon themselves to be 16 years old. Their mental age would be 14 (if they had passed the test for 14 year-olds) plus five months for every problem they performed in the test for average adults. This test had six problems. They could also add on six months for every problem of the six problems in the test for superior adults. The maximum mental age which could be reached was 19.5.

In association with the 1916 revision of the Binet test, Terman published a classification of human beings:

IQ	*Classification*
Over 140	Genius or near genius
120–140	Very superior intelligence
110–119	Superior intelligence
90–109	Normal or average intelligence
80–89	Dullness
70–79	Borderline deficiency
Under 70	Definite feeble-mindedness

In 1937 Stanford-Binet 16 was revised by Professor Terman, now in collaboration with Dr. Maude A. Merrill. also associated with Stanford University.

Terman and Merrill started with the 90 problems from the 1916 test plus several hundred more problems collected in various ways. Of these most were dropped as work went on and the 1937 Stanford-Binet finally contained 258 problems divided into two parallel test series with 129 problems in each. So there was an increase of 40 problems compared to 1916.

From thousands of problems, all but 410 were first sifted out. The sifting was based on assessment of validity, but also on practical considerations such as ease of scoring, etc. Validity was principally judged by the criterion of increase in the percentages of right answers from experimental groups of more than 600 children from one age to the next.

When this sifting process was completed Terman and his colleagues assembled 3,184 children of all ages up to 18. (After that age Terman believed that intelligence could no longer increase.) The 3,184 were all native Americans and members of the white race. No particular nationality group was excluded. They were very carefully selected in order to represent a social cross-section of the white American community.

The 3,184 children were given the 410 problems, i.e. those of them which were suitable with regard to degree of difficulty, and then work began on a massive scale. All the answers, and quantities of other data were recorded on punch-cards and thousands of correlation coefficients were calculated. In all, 152 of the 410 problems were excluded, among other reasons because they did not correlate highly with the test as a whole. This meant that whatever the original 1916 test measured, the fresh problems were guaranteed, broadly speaking, to measure the same thing. Terman and Merrill write: "In the case of Form L" (one of the two parallel test series L and M) "six successive revisions were necessary to accomplish the result, but once this had been done for Form L, it was possible to achieve at once an equally good result with Form M by arranging its tests so as to match those of Form L at each age level with respect to difficulty, validity and shape of curve of percents passing by age. The standardization procedure involved not only the shifting about of tests . . . but also . . . modifying the standard of scoring to make a given test easier or harder so as to make it fit a given age level, etc."

The difference between girls and boys in the percentage of right answers to a problem was compensated by shifting problems in which the opposite variations were found from one form to the other.

Terman and Merrill wrote a large book about how they designed the 1937 test. It is interesting to note that it says nothing about validity criteria other than purely statistical ones. Apparently Terman's definition of intelligence was: that which in a certain statistical way distinguishes the capacity of children to solve problems of the type which Binet used in his test. But Terman shows that he is aware that other psychologists have different criteria, even if he considers them wrong: "The hypercritical . . . will continue to oppose its use on grounds which are largely irrelevant: because it is often misunderstood or misused; because it is influenced by environment as well as by endowment; because it does not measure social adaptability, because it does not always predict accurately success in school or success in life, etc."

Terman considered, strangely enough, that it was unlikely that intelligence really was normally distributed: ". . . that in an unselected population the distribution of intelligence follows strictly the normal curve . . . may or may not be true. There are biological characters for which it is not true and intelligence may conceivably be one of them. The question [how is intelligence distributed?] could be answered for intelligence if we had an equal unit scale to begin with, but we are in the unfortunate position of having to assume the answer in advance in order to derive the equal-unit scale."

He nevertheless put a good deal of work into arranging his test problems so that their results were distributed along the normal curve.

The 1960 revision of Stanford-Binet consisted in getting rid of the two parallel tests. The problems which continued to sort people out effectively were kept and the rest thrown out. As in 1937, the problems which did not correlate highly with the test as a whole were also rejected. Finally a good many problems which had become out of date were dropped. For instance, in the 1930s 69% of 3 year-olds in the standardization group were able to recognize and name five out of six objects, consisting of small-scale models of a shoe, a watch, a telephone, a flag, a pocket-knife and a stove. In the 1950s only 11% of the children with a mental age of 3 could do this, because the telephone was a '30s style apparatus, the stove was old-fashioned, etc. In other words, the objects were quite strange to most children of the 1950s.

Intelligence and intelligence factors

Factor analysis is the name given to a certain method of dealing with correlation coefficients. The method is partly based on evaluation. The Swedish psychologist Jan Gästrin writes: "Factor anlysis will not give us an objective answer to what intelligence is. The decisive steps are marked by subjective interpretations."

If we give a group of people an intelligence test consisting of ten sub-tests we can calculate the correlation between each sub-test and any other sub-test. In all we will obtain 45 different correlation coefficients. These can be put on a chart called a correlation matrix. The basis of all factor analysis is a correlation matrix of the type shown in Figure 48.

In the correlation matrix in Figure 48 the correlations between opposite word problems, word definition problems and completing incomplete sentence problems are higher than the other correlations. This means that the results of these three sub-tests have something in common. It is assumed that these three sub-tests measure the same part of the intelligence, and this part is called an intelligence factor. Since all three sub-tests are concerned with words, the factor is called the verbal factor.

Instead of simply looking at the correlation matrix and seeing which tests have the highest correlations with each

Fig. 48. Correlation matrix (the nought in the correlation coefficients has been omitted).

	OPPOSITE WORD PROBLEMS	ARITHMETICAL PROBLEMS	WORD DEFINITION PROBLEMS	BLOCK COUNTING PROBLEMS	JIGSAW PUZZLES	COMPLETING INCOMPLETE SENTENCES	LOGICAL PROBLEMS	INFORMATION	RAVEN'S MATRICES	DISTINGUISHING BET. GEOMETRICAL FIGS.
OPPOSITE WORD PROBLEMS										
ARITHMETICAL PROBLEMS	.23									
WORD DEFINITION PROBLEMS	.79	.11								
BLOCK COUNTING PROBLEMS	.13	.52	.09							
JIGSAW PUZZLES	.32	.14	.20	.23						
COMPLETING INCOMPLETE SENTENCES	.82	.31	.75	.18	.35					
LOGICAL PROBLEMS	.27	.41	.21	.16	.19	.09				
INFORMATION	.41	.21	.43	.17	.11	.31	.12			
RAVEN'S MATRICES	.37	.18	.14	.31	.14	.29	.28	.11		
DISTINGUISHING BET. GEOMETRICAL FIGS.	.12	.25	.12	.42	.27	.17	.13	.19	.31	

other and investigating the relationships of these tests to the other tests in the matrix, one can programme a computer to do the same thing. Then if we feed the figures from the correlation matrix into the computer, after a time we get out information as to the tests which have common factors, the relationship of these factors to each other and their relationship to various sub-tests in the correlation matrix. If we give the computer different programmes we shall obtain information covering different factors and different relationships between them. This is true even if we feed the computer with the same correlation matrix. The operations which are jointly called factor analysis give different results. They function like computers programmed in different ways.

The conclusions which psychologists draw from the results of factor analyses are of this type:

1. Intelligence consists of this and that factor, plus this or that many secondary factors, plus this and that tertiary factor.
2. This test measures the factor *f* better than that one.

Conclusions of this type have led many readers of modern psychology textbooks to believe that the factors are natural, almost biological units. In fact, different psychologists using different factor-analytical methods produce different intelligence factors on the basis of the same correlation matrix. It should also be noted that the intelligence factors are factors of what is measured by intelligence tests and nothing else. This means that intelligence factors are parts of the test to the same extent as they are parts of something within the human being.

When factor analysis began to be applied to intelligence tests it was found that all sub-tests correlated with each other to a greater or less extent. When the common factor was excluded it was found that each sub-test also had a specific factor. So there were two types of factors: the general intelligence factor, the *g* factor, and also a factor which was specific for each sub-test, an arithmetical factor for arithmetical problems, a word factor for verbal problems, etc. This two-factor theory was introduced by Charles Spearman. English psychologists have followed his lead and consider that intelligence consists of the *g* factor plus some secondary factors such as memory, ability to distinguish between symbols, ability to manipulate words, etc.

American psychologists, on the other hand, came to use factor-analytical methods and evaluations which resulted in the conclusion that intelligence consists of seven different factors: a verbal factor, a flexibility factor, a logical factor, a spatial preception factor, a numerical factor, a discrimination factor and an analogy factor. Besides this, intelligence consists of several other secondary factors, plus one or a few tertiary factors of a general nature, which correspond to Spearman's *g* factor.

Broadly speaking, factor analysis will produce as many factors as there are different types of problem in the test. The factors reflect the types of behaviour to which different types of symbols and symbol rules in the various types of problem give rise. The factors are therefore probably more an expression of common characteristics of the problems and common characteristics of the reactions to which the various problems give rise in different people in test situations than an expression of an inherent common structure of the brain or "thinking". If the Stanford-Binet test is translated into Chinese, which is a picture language with symbols and symbol rules quite different from those of Western languages, and if the result of the Chinese Stanford-Binet test is factor-analyzed with current methods, one would probably find quite different factors, with different inherent relationships, from those found in England and the USA based on the "same" test. (Of course the test would not be the same test because it would have been converted into a completely different symbol system.)

I will develop this with one or two examples. In Chinese the same term is used to describe the things we call "cause" and "motive". Just suppose that a question in the test we are to translate into Chinese runs: "What is the difference between 'cause' and 'motive'?"

The Stanford-Binet test contains problems of this type:

"A rabbit is timid, a lion is . . .?" If we gave a group of Swedish children this test, plus a few other tests including the test of the capacity to find words which rhymed with each other, we would probably find a very high correlation in our correlation matrix between this type of Stanford-Binet problem and the rhyming test. People who found it easy to give the correct opposite word in this type of Stanford-Binet problem: "Timid–brave", "swim–fly", "deciduous–evergreen", etc., would probably have grown up in a nursery atmosphere and home environment which made them cleverer than other people at thinking of the rhymes: "Heart, smart, mud, blood, steel, meal", etc.

When we translate this type of Stanford-Binet problem into Chinese it will probably become incredibly difficult or perhaps impossible to reproduce "A rabbit is timid, a lion is . . . ?" in such a way that an ordinary, well-known pair of opposites in Chinese can be found to replace "timid–brave". But even if this translation succeeds, it is not certain that rhymes can be found in our sense of the word in the Chinese language. And even if there is a way of rhyming words in Chinese which corresponds to our own, it is not certain that the ability to rhyme is associated for the Chinese with the ability to complete pairs of opposites.

We would probably be determining quite different factors in the Chinese intelligence from those we discover by factor analysis in Europe and America.

The usefulness of factor analysis consists in its ability to detect connections between problems and modes of reaction which would be difficult to detect by simply looking at the correlation matrices. These methods can also be used to acquire more knowledge of the basic structure of the symbol system of a particular culture and of the way of mentally manipulating symbols practised by large ethnic groups.

But factor analysis can tell us little about an individual's mental structure. We are completely dependent on subjective assessments of another person's mental activities.

Factor-analytical methods have given rise to a special validity concept: factorial validity. By means of factor analysis we find that a collection of problems has something in common, which means that people who can perform one of the tasks in the collection well can generally also perform the other tasks in the collection equally well. When we know what tasks have something in common in this way, we have to find out in what the common factor consists. And here we come across one of the most important of the subjective evaluations involved. Professor Leona E. Tyler writes: "There is often room for a considerable difference of opinion as to just what it is that several tests have in common." L. L. Thurstone, the best known of American factor analysts, describes how, in doubtful cases, he asks a number of experimental subjects how they have got on in solving a problem and then asks them to describe any tricks and dodges they have discovered when doing the test. These informal reports are very enlightening. A little practice of this kind makes the importance of introspection very clear by comparison with the superficial appreciation of the factorial composition of a test simply by a cursory inspection of its content.

Once we have discovered what the highly correlated results of test situations have in common, we give them a name. If the highly correlated problems all involve arithmetic, for instance, we call this factor the numerical factor. By giving it a name one has so to speak *created a new attribute* in human beings, an attribute which is part of the greater attribute of intelligence.

Now we can distinguish the problems which measure this common numerical factor from the rest of the problems and assemble them in one particular test. If we give this new numerical-factor test to a large group of people we can work out the correlation of every single problem in the test with the test as a whole. We will then find that many problems have a lower correlation with the test as a whole than others. These problems are then said to have low factorial validity. They are not very highly loaded on the numerical factor.

When the numerical factor has been detected and isolated and when we have a test which is highly loaded on this factor, we try to design new problems which have high factorial validity as regards the numerical factor, thus 'cultivating' problems of a certain type. In this case, we shall be producing arithmetical problems which in essential respects resemble the problems in the test.

In much modern intelligence research there is little talk of intelligence in general. We speak of the *v* factor, the *n* factor, etc., on American lines.

The methods and character of factor analysis mean that the intelligence factors mentioned will be disappearing, and being replaced by others, if factor analytical methods remain constant while the symbol system and reactions to it are changing. If we use L. L. Thurstone's factor analysis of correlation coefficients from a number of arithmetical tests given to young Swedish schoolchildren, we will probably find one or more new factors. This is because mathematical teaching has recently been reconstructed so that set theory is the basis of understanding of the mathematical symbol system. Set theory is a branch of the fundamental Western way of thinking which is called logic. It entails a new and slightly different way of reacting to mathematical problems from the older arithmetical methods.

Previously schoolchildren had to learn a certain series of figures and the rules for manipulating them. For instance, they had to learn the natural numbers 1, 2, 3, 4, etc., and the + symbol and the – symbol. When they knew these symbols and the rules governing them, they had to learn a new series of figures, for instance fractions and their related symbols: the division sign and the multiplication sign. Then they had to work on these symbols until they had learned the rules of the system. This method of teaching enabled children to learn to carry out a good many ordinary arithmetical operations mechanically, without really understanding that mathematics is a game of symbols, with rules for moving and changing the symbols. If a child had some basic misunderstanding about the section on natural numbers and plus and minus, he could still manage his arithmetic in the next class and perhaps the next one as well. But when he got to high school things got more and more laborious and finally he had no idea what was going on. He was not "mathematically gifted." At the high school stage it was too late to go back and work out which of the most elementary symbols or rules had not been understood.

The introduction of set theory at the very earliest stage involves a completely different way of approaching mathematics. It is now clear from the very beginning that mathematics is a game of symbols which denote quantities, with the rules governing them. Instead of jumping straight into the mathematical game and learning the symbols and rules for a little bit of the great game, the child is now able to survey the great game itself first and then the most fundamental rules for the manipulation of the set symbols. By this process he is given simultaneously a greater distance from the game and a greater ability to play it.

Professor Tyler writes of factor analysis: "Thus the factor analyst can only say, 'This is *one* combination of traits which would serve to account for the relationships we have found between these tests.' Another research worker may propose another set of traits based on equally sound mathematical and psychological reasons, which will account for the relationship equally well. The choice between them must be made on the grounds of simplicity, usefulness and congruence with the whole body of psychological knowledge. This is why there are so many arguments among factor analysts . . ."

POSTSCRIPT

It is no exaggeration to say that most people believe that the attribute measured by intelligence tests is inherited and that there is no point in trying to understand what cannot be understood. You can't change your nature. To quote a proverb used in Stanford-Binet: "You can't make a silk purse out of a sow's ear".

It is unlikely that many of these people will be reading this book, but those who have done so will probably have realized that they have been the victims of indoctrination. It is probable to the point of certainty that the variations in people's intelligence are not genetically based, in other words, that the variations between different individuals in what is measured by intelligence tests do not depend to any great extent on inherent differences in the structure of the brain. The variations in IQ are almost wholly dependent on differences in the individual's experience from the beginning of his life onwards, including what happens between conception and birth.

For four or five years I taught in most classes of Swedish schools: junior secondary schools, girls' schools, sixth form colleges, state high schools, comprehensives and evening classes. All the psychology books I have seen have given teachers the impression that inherent differences in the capacity of the brain are of decisive importance to the level of the IQ. Even in the most modern textbooks for sixth form level this indoctrination persists. In psychology books for university level the uncertainty of twin diagnosis is not mentioned at all.

I hope with this book to have dealt a deathblow to the intelligence myth. I also hope that the book will contribute to a more consistent use of a word which is so loaded with value judgements.

The book is written with the aim of giving an interested public a broad, and preferably easily readable survey of a very complex subject.

The theoretical basis of those parts of the book which concern the concept of intelligence are a sketch for a work on the concept of intelligence from the semiotic aspect.

I regard it as extremely important to analyse the use of current terms (frequently used in connection with social intervention in the form of isolation, punishment, care, etc.) such as mental illness, intelligence, schizophrenia and hallucination. And by analysis of use I do not mean getting hold of an established theory in which the concept is included and finding out if the concept is used in the theory in a logically reasonable way. I mean an unprejudiced analysis, based on how the concept is actually used in society: in school books, as motivation for social intervention, in daily speech, in various sciences.

An analysis of this nature includes studies of data from other disciplines, in as much as such data are evoked in the use of the concept in question. In discussing the concept of intelligence, I have had to cross the boundaries between theoretical philosophy and several other disciplines: the history of science and ideas, sociology, pedagogy, psychology, genetics and anthropology.

Malmö, 7th April 1970 Carl G. Liungman

In the second edition a number of minor linguistic obscurities have been corrected.

Malmö, 3rd February 1972 Carl G. Liungman

For the English language edition of the book a number of less happy formulations have been omitted, some coefficients have been included in more precise form and a new chapter has been added (From Test Results to Conclusions on Heredity and Environment). Some recent English and American works have been added to the bibliography.

Kristianstad, 13th June 1974 Carl G. Liungman

BIBLIOGRAPHY

Anne Anastasi: *Psychological Testing*. 2nd ed., MacMillan Company. New York 1963.

– *Differential Psychology: Individual and Group Differences in Behavior*. 3rd ed., MacMillan Company. New York 1958.

– "The Concept of Validity in the Interpretation of Test Scores". *Readings in Educational and Psychological Measurement.* Editors: Clinton I. Chase and Glenn H. Ludlow. Houghton Mifflin Company. Boston 1966.

Anne Anastasi (editor): *Individual Differences*. John Wiley & Sons Inc. USA 1965.

C. Arnold Anderson: "Access to Higher Education and Economic Development". *Education Economy and Society – a Reader in the Sociology of Education*. Editors: A. H. Hasley, Jean Floud and C. A. Anderson. Free Press. New York 1969.

William H. Angoff: "Scales with Nonmeaningful Origins and Units of Measurement", see Anne Anastasi: "The Concept of Validity . . ."

Louise Behrens Apperson and W. George McAdov Jr: "Parental Factors in the Childhood of Homosexuals". *Journal of Abnormal Psychology*, 1968 Vol. 75 No 3.

Lotte Bailyn: "Mass Media and Children: A Study of Exposure Habits and Cognition". *Psychological Monographs*, 1959 No 471.

Alfred L. Baldwin: *Behavior and Development in Childhood*. Holt, Rinehart and Winston. New York 1965.

Howard S. Becker: "Schools and Systems of Stratification", see C. Arnold Anderson.

Gösta W. Berglund: "A Note on Intelligence and Seasons of Birth". *British Journal of Psychology,* 1967 No 58: 1 and 2.

Basil Bernstein: "Social Class and Linguistic Development: A Theory of Social Learning", see C. Arnold Anderson.

A. Binet et Th. Simon: "Méthodes nouvelles pur le diagnostic du niveau intellectuel des anormaux". *Année psychol.*, 1905 No 11.

Josph B. Birdsell: "Some Environmental and Cultural Factors Influencing the Structuring of Australian Aboriginal Populations". *Human Ecology – Collected Readings*. Editor: Jack B. Bresler. Addison-Wesley Publishing Company Inc. USA 1966.

John Blomqvist: *Elevers skolanpassning*. Almqvist & Wiksell. Uppsala 1969.

Robert Borger and E. M. Seaborne: *The Psychology of Learning*, Penguin Books Ltd.

Jack B. Bresler: "Maternal Height and the Prevalence of Stillbirth", see Joseph B. Birdsell.

D. E. Broadbent: "Response to Stress in Military and Other Situations". *Readings in Psychology*. Editor: John Cohen. George Allen & Unwin Ltd. London 1964.

Edmund des Brunner and Sloan Wayland: "Occupation and Education", see C. Arnold Anderson.

Cyril Burt: "The Evidence for the Concept of Intelligence". *British Journal of Educational Psychology*, 1955 Vol. 25.

– "Francis Galton and his Contributions to Psychology", see D. E. Broadbent.

D. Campbell: "The Psychological Effects of Cerebral Electroshock". *Handbook of Abnormal Psychology – an Experimental Approach*. Editor: H. J. Eysenck. Pitman Medical Publishing Company Ltd. London 1968.

C. O. Carter: *Human Heredity*. Penguin Books Ltd.

Norman Corah et al.: "Effects of Perinatal Anoxia after Seven Years". *Psychological Monographs*, 1965 No 596.

Lee J. Cronbach and Paul E. Meehl: "Construct Validity in Psychological Tests", see Anne Anastasi: "The Concept of Validity . . ."

Herbert J. Cross: "The Relation of Parental Training Conditions to Conceptual Level in Adolescent Boys". *Journal of Personality* 1966 March-Dec.

J. F. Dishiell: *Fundamentals of General Psychology*. 3rd ed., Houghton Mifflin Company. Cambridge, Mass.

Lois-Ellin Datta: "Draw-a-Person Test as a Measure of Intelligence in Preschool Children from Very Low Income Families". *Journal of Consult. Psych.*, 1967 Vol. 31 No 6.

James E. Deese: *The Psychology of Learning*. McGraw-Hill Book Company, Inc.

John F. Defee Jr and Philip Himelstein: "Children's Fears in a Dental Situation as a Function of Birth Order". *Journal of Genetic Psychology*, Dec. 1969 Vol. 115: 2.

Richard DeMille: "Intellect after Lobotomy in Schizophrenia. A Factor-analytic Study". *Psychological Monographs*. 1962 No 535.

Wayne Dennis: "Infant Development under Environmental Handicap". *Psychological Monographs*, 1957 No 436.

J. B. Deregowski: "Difficulties in Pictorial Depth Perception in Africa". *British Journal of Psychology*, August 1968.

Martin Deutsch: "Happenings on the Way Back to the Forum: Social Science, IQ, and Race Differences Revisited". *Harvard Educational Review*, 1969 Vol. 39 No 3.

– "Equal Educational Opportunity". *Harvard Educational Review* Cambridge Mass. 1969.

H. J. Eysenck: "Classification and the Problem of Diagnosis", see D. Campbell.

– *Fact and Fiction in Psychology*, Penguin Books Ltd, 1968.

– *The Inequality of Man*. Temple Smith, London 1973.

F. H. Farley: "Season of Birth, Intelligence and Personality". *British Journal of Psychology*, August 1968.

F. S. Fehr: "Critique of Hereditarian Account of Intelligence and Contrary Findings: A Reply to Jensen". *Harvard Educational Review*, 1969 Vol. 39 No 3.

Cyril M. Franks: "Conditioning and Abnormal Behaviour", see D. Campbell.

Frank S. Freeman: *Theory and Practice of Psychological Testing.* 3rd ed., Holt, Rinehart and Winston. USA 1964.

W. D. Furneaux: "Intellectual Abilities and Problem-solving Behaviour", see D. Campbell.

Martin Gardner: *The Ambidextrous Universe*. Basic Books Inc. New York 1964.

Henry E. Garrett: *Great Experiments in Psychology.* 3rd ed., Bonniers and Appletons-Century-Crafts Inc. USA 1955.

John T. Gentry: "An Epidemiological Study of Congenital Malformations in New York State", see Joseph B. Birdsell.

Warren R. Good: "Misconceptions about Intelligence Testing", see Anne Anastasi: "The Concept of Validity . . ."

Frances K. Graham et al.: "Development 3 Years after Perinatal Anoxia and Other Potent Dangers Newborn Experience". *Psychological Monographs*, 1962 No 522.

Robert J. Havighurst: "Education and Social Mobility in Four Societies", see C. Arnold Anderson.

L. T. Hilliard and Brian H. Kirkman: *Mental Deficiency* J & A. Churchill Ltd. London 1957.

Paul H. Hoch and Joseph Zubin (editors): *Relation of Psychological Tests to Psychiatry*. Grune & Stratton. New York 1952.

Marjorie P. Honzik, Jean W. Macfarlane and Lucile Allen: "The Stability of Mental Test Performance between Two and Eighteen Years", see Anne Anastasi: "The Concept of Validity . . ."

Alvin R. Howard: "A Fifteen-Year Follow-up with the Wechsler Memory Scale". *Journal of Consult. Psych.*, 1966 Vol. 30 No. 2.

Ian M. Hunter: "The Organization of Memory", see D. E. Broadbent.

Torsten Husén: "Abilities of Twins". *Scandinavian Journal of Psychology*, Vol. 1 1960.

– *Begåvning och miljö*. Hugo Gebers Förlag. Stockholm 1951.

Sauli Häkkinen and Isto Ruoppila: "The Effect of Performance Time and Retesting upon the Factor Structure of Intelligence Tests". *Scandinavian Journal of Psychology*, 1960 Vol. 1.

Bärbel Inhelder: "Some Aspects of Piaget's Genetic Approach to Cognition", see D. E. Broadbent.

Marie Jahoda: *Race Relations and Mental Health*. Unesco. Belgium 1960.

Christopher Jencks: *Inequality – A Reassessment of the Effect of Family and Schooling in America*. Allen Lane, London 1973.

S. W. Johnson and R. J. Stiggins: "A Cross-Cultural Study of Values and Needs". *Acta Psychologica*, 1969 Vol 31.

H. Gwynne Jones: "Learning and Abnormal Behaviour", see D. Campbell.

Hilda Knobloch and Benjamin Pasamanick: "Seasonal Variation in the Births of the Mentally Deficient", see Joseph B. Birdsell.
David Krech and Richard S. Crutchfield: *Elements of Psychology.* Alfred A. Knopf. New York 1958.
Roger T. Lennon: "A Comparison of Results of Three Intelligence Tests", see Anne Anastasi: "The Concept of Validity . . ."
Elsie E. Lessing: "Racial Differences in Indices for Ego Functioning Relevant to Academic Achievement". *Journal of Genetic Psychology*, Dec. 1969 Vol. 115: 2.
C. A. Mace: "Homeostasis, Needs and Values", see D. E. Broadbent.
David Magnusson: "Some Personality Tests Applied on Identical Twins". *Scandinavian Journal of Psychology*, 1960 Vol. 1.
Jeanne Cummins Mellinger and and Ernest A. Haggard: "Personality. Intellectual and Achievement Patterns in Gifted Children". *Psychological Monographs*, 1959 No 483.
Patrick Meredith: "Models, Meanings and Men" see D. E. Broadbent.
V. Meyer: "Psychological Effects of Brian Damage", see D. Campbell.
Ian W. Monie: "Influence of the Environment of the Unborn", see Joseph B. Birdsell.
Clifford T. Morgan and Richard A. King: *Introduction to Psychology.* 3rd ed., McGraw-Hill Book Company.
Charles I. Mosier: "A Critical Examination of the Concept of Face Validity", see Anne Anastasi: "The Concept of Validity . . ."
Norman L. Munn: *Psychology – the Fundamentals of Human Adjustment.* 4th ed., George G. Harrap & Co. Ltd. USA 1961.
John Nisbet: "Family Environment and Intelligence" see C. Arnold Anderson.
Magne Nyborg: *Noen teoretiske synspunkter på mental retardasjon. En litteraturstudie.* Universitetsforlaget. Oslo 1969.
N. O'Connor and Cyr. M. Franks: "Childhood Upbringing and Other Environmental Factors", see D. Campbell.
R. W. Payne: "Cognitive Abnormalities", see D. Campbell.
Elizabeth Peal and Wallace E. Lambert: "The Relation of Bilingualism to Intelligence". *Psychological Monographs*, 1962 No 546.
Jean Piaget: *Intelligensens psykologi.* Natur & Kultur, Stockholm 1951.
L. Rees: "Constitutional Factors and Abnormal Behaviour", see D. Campbell.
J. A. Fraser Roberts: *An Introduction to Medical Genetics.* 3rd ed., Oxford University Press. London 1965.
Peter H. Rossi: "Social Factors in Academic Achievement: a Brief Review", see C. Arnold Anderson.
Kenneth Ruig et al.: "The Relationship of Birth Order to Self-evaluation, Anxiety Reduction, and Susceptibility to Emotional Contagion". *Psychological Monographs*, 1965 No 603.
K. Warner Schaie: "Rigidity-Flexibility and Intelligence: A Cross-sectional Study of the Adult Life Span from 20 to 70 Years". *Psychological Monographs*, 1958 No 462.

W. J. Schull and J. V. Neal: *The Effects of Inbreeding on Japanese Children*. Harper & Row, New York 1965.

James Shields: *Monozygotic Twins Brought Up Apart and Brought Up Together*. Oxford University Press, London 1962.

J. Shields and E. Slater: "Heredity and Psychological Abnormality"; see D. Campbell.

M. Skodak and H. M. Skeels: "A Final Follow-up Study of One Hundred Adopted Children", *The Journal of Genetic Psychology*, 1949, Vol. 75, pp. 85–125.

Kaj Spelling: *Intelligens og tænkning*. Berlingske Forlag. Copenhagen 1968.

– *Miljoets indflydelse på intelligensudviklingen*. Nyt Nordisk Forlag Arnold Busck, Copenhagen, 1963.

Abraham A. Spevack and Milton D. Suboski: "Retrograde Effects of Electroconvulsive Shock on Learned Response', *Psychological Bulletin,* 1969 Vol. 72 No 1.

Curt Stein: *Principles of Human Genetics.* 2nd ed., W. H. Freeman & Company, San Francisco and London. USA 1960.

L. M. Terman and M. A. Merrill: *Measuring Intelligence*. Houghton Mifflin, Boston 1937.

Robert L. Thorndike and Elizabeth Hagen: *Measurement and Evaluation in Psychology and Education*. John Wiley & Sons Inc. New York, London 1961.

Louis P. Thorpe: *Child Psychology and Development*. 2nd ed., The Ronald Press Company. New York 1955.

Harald Torpe: *Intelligensforskning og intelligensprøver*. 3rd ed., J. H. Schulz Forlag. Copenhagen, 1964.

A. F. Tredgold: *A Text-book of Mental Deficiency (Amentia)*. Ballière, Tindall and Cox. London 1952.

Leona E. Tyler: *The Psychology of Human Differences.* 3rd ed., Appleton-Century-Crofts. New York 1965.

– *Tests and Measurements.* Prentice-Hall Inc. USA 1964.

Jack A. Vernon: *Inside the Black Room–Studies in Sensory Deprivation.* Penguin Books. Great Britain 1966.

P. E. Vernon: "The Psychology of Intelligence and G", see D. E. Broadbent.

Morris Weitman: "More than one kind of Authoritarian". *Journal of Personality*, Vol. XXX March–Dec. 1962.

R. Willet: "The Effects of Psychosurgical Procedures on Behaviour", see D. Campbell.

Stephen Wiseman (editor): *Intelligence and Ability*. Penguin Books. Bungay 1967.

J. R. Wittenborn: "A Study of Adoptive Children". *Psychological Monographs*, 1956 Nos 408, 409, 410.

Dael Wolfe: "Educational Opportunity Measured Intelligence and Social Background", see C. Arnold Anderson.

D. H. Young and T. D. Glover: "Temperature and the Production of Spermatozoa", see Joseph B. Birdsell.

James Youniss and Han G. Furth: "The Role of Language and Experience on the Use of Logical Symbols". *British Journal of Psychology*, 1967 No 58: 3 and 4.

Hans Olof Åkesson: *Medicinsk genetik*. Almqvist & Wiksell. Stockholm 1968.

INDEX